RAND MCNALLY

GOODE'S

ATLAS OF Human Geography

distributed by
WILEY John Wiley & Sons, Inc.

Howard Veregin, Ph.D., Editor

Editorial Advisory Board

Byron Augustin, D.A., Texas State University-San Marcos

Joshua Comenetz, Ph.D., University of Florida

Francis Galgano, Ph.D., United States Military Academy

Sallie A. Marston, Ph.D., University of Arizona

Virginia Thompson, Ph.D., Towson University

Abridgement of
21ST Edition

Working together to bring you the best in geography education

Few publishers can claim as rich a history as John Wiley & Sons, Inc. (publishers since 1807) and Rand McNally & Company (publishers since 1856). Even fewer can claim as long-standing a commitment to geographic education.

Wiley's partnership with the geographic community began at the very beginning of the 20th century with the publication of textbooks on surveying. Rand McNally's partnership began even earlier, with the publication of the first Rand McNally maps in 1872. Since then, both companies have worked in parallel to help students visualize spatial relationships and appreciate the earth's dynamic landscapes and diverse cultures.

Now these two publishers have combined their efforts to bring you this new atlas, which represents the very best in educational resources for geography.

Based on the 21st edition of the *Goode's World Atlas*, the *Goode's Atlas of Human Geography* features:

- An emphasis on map accuracy and legibility, and the mixture of maps of different types and scales to facilitate interpretation of geographic phenomena.

- World, continental, and regional population density maps, which have been created using LandScan, a digital population database developed using satellite and computer-mapping technology.

- Graphs accompanying many of the maps, to show important statistical information, trends over time, and relationships between variables.

- Maps and graphs that have been updated, based on the most current available data in accordance with the high standards and quality that have always been a defining feature of the *Goode's World Atlas*.

Wiley and Rand McNally are currently offering seven new course-specific atlases, which can be packaged with any of Wiley's best-selling textbooks, or sold separately as stand-alones. These atlases include:

Rand McNally Goode's Atlas of Political Geography	0-471-70694-9
Rand McNally Goode's Atlas of Latin America	0-471-70697-3
Rand McNally Goode's Atlas of North America	0-471-70696-5
Rand McNally Goode's Atlas of Asia	0-471-70699-X
Rand McNally Goode's Atlas of Urban Geography	0-471-70695-7
Rand McNally Goode's Atlas of Physical Geography	0-471-70693-0
Rand McNally Goode's Atlas of Human Geography	0-471-70692-2

This book was set by GGS Book Services and printed and bound by Walsworth Press. The cover was printed by Phoenix Color.

To order books or for customer service please, call 1-800-CALL WILEY (225-5945).

ISBN 0471-70692-2

Printed in the United States

10 9 8 7 6 5 4 3 2 1

Table of Contents

Introduction

iv   Basic Earth Properties

v   Map Scale

vi   Map Projections

vii   Map Projections Used in *Goode's Atlas of Human Geography*

ix   Thematic Maps in *Goode's Atlas of Human Geography*

Maps

1   Map Legend

2, 3   Political

4, 5   Physical

6, 7   Population Density

8   Birth Rate / Death Rate

9   Natural Increase / Life Expectancy

10   Gross Domestic Product / Literacy

11   Languages / Religion

12   Urbanized Population / Nutrition

13   Physicians / HIV Infection

14, 15   Major Agricultural Regions
    INSETS: Probable Origins of Cultivated Plants

16   Wheat / Tea, Rye

17   Maize (Corn) / Coffee, Oats

18   Barley, Cocoa Beans / Rice, Millet and Grain Sorghum

19   Potatoes, Cassava / Cane Sugar, Beet Sugar

20   Fruits / Tobacco, Fisheries

21   Vegetable Oils

22   Natural Fibers, Rubber / Beer and Wine

23   Cattle / Pigs

24   Sheep / Poultry

25   Food Aid / Fertilizer Use

26, 27   Forested Lands and Products

28   Copper / Tin, Bauxite

29   Lead / Zinc

30   Iron Ore and Ferroalloys

31   Steel / Precious Metals

32   Nuclear and Geothermal Power / Hydroelectricity

33   Energy Production / Energy Consumption

34, 35   Mineral Fuels

36   Exports / Imports

37   Military Power / Women's Rights

38   Political and Military Alliances / Economic Alliances

39   Refugees / Major Conflicts 1990–2003

40   North America—Energy / Water Resources / Natural Hazards / Landforms

41   North America—Annual Precipitation / Vegetation / Population / Economic, Minerals

42   North America—Environments *Scale 1:36,000,000*

43   United States and Canada—Transportation / Westward Expansion

44, 45   United States and Canada—Physiography *Scale 1:12,000,000*

46   United States and Canada—Precipitation / Glaciation

47   United States and Canada—Climate

48, 49   United States and Canada—Natural Vegetation

50, 51   United States and Canada—Agriculture

52   United States—Water Resources

53   United States and Canada—Population

54   United States—Demographics

55   United States—Demographics

56   United States—Demographics

57   United States and Canada—Labor Structure / Value Added by Manufacturing

58, 59   United States and Canada—Federal Lands and Interstate Highways

60   North America—Political *Scale 1:40,000,000*

61   North America—Physical *Scale 1:40,000,000*

62   South America—Energy / Peoples / Natural Hazards / Landforms

63   South America—Annual Precipitation / Vegetation / Population / Economic, Minerals

64, 65   Europe—Languages *Scale 1:16,500,000*

66   Middle America—Ethnic / Political

67   Northern Eurasia—Ethnic
    Middle East—Ethnic

68   Africa—Political Change / Peoples / Natural Hazards / Landforms

Tables and Indexes

69   World Political Information Table

74   World Demographic Table

76   World Agriculture Table

78   World Economic Table

80   World Environment Table

82   World Comparisons

83   Principal Cities of the World

84   Glossary of Foreign Geographical Terms

85   Abbreviations of Geographical Names and Terms
    Pronunciation of Geographical Names

87   Pronouncing Index

93   Subject Index

94   Sources

Introduction

Basic Earth Properties

The subject matter of **geography** includes people, landforms, climate, and all the other physical and human phenomena that make up the earth's environments and give unique character to different places. Geographers construct maps to visualize the **spatial distributions** of these phenomena: that is, how the phenomena vary over geographic space. Maps help geographers understand and explain phenomena and their interactions.

To better understand how maps portray geographic distributions, it is helpful to have an understanding of the basic properties of the earth.

The earth is essentially **spherical** in shape. Two basic reference points — the **North and South Poles** — mark the locations of the earth's axis of rotation. Equidistant between the two poles and encircling the earth is the **equator**. The equator divides the earth into two halves, called the **northern and southern hemispheres**. (See the figures to the right.)

Latitude and longitude are used to identify the locations of features on the earth's surface. They are measured in degrees, minutes and seconds. There are 60 minutes in a degree and 60 seconds in a minute. Latitude is the angle north or south of the equator. The symbols °, ', and " represent degrees, minutes and seconds, respectively. The N means north of the equator. For latitudes south of the equator, S is used. For example, the Rand McNally head office in Skokie, Illinois, is located at 42°1'51" N. The minimum latitude of 0° occurs at the equator. The maximum latitudes of 90° N and 90° S occur at the North and South Poles.

A **line of latitude** is a line connecting all points on the earth having the same latitude. Lines of latitude are also called **parallels**, as they run parallel to each other. Two parallels of special importance are the **Tropic of Cancer** and the **Tropic of Capricorn**, at approximately 23°30' N and S respectively. This angle coincides with the inclination of the earth's axis relative to its orbital plane around the sun. These tropics are the lines of latitude where the noon sun is directly overhead on the solstices. (See figure on page 66.) Two other important parallels are the **Arctic Circle** and the **Antarctic Circle**, at approximately 66°30' N and S respectively. These lines mark the most northerly and southerly points at which the sun can be seen on the solstices.

While latitude measures locations in a north-south direction, longitude measures them east-west. Longitude is the angle east or west of the **Prime Meridian**. A **meridian** is a line of longitude, a straight line extending from the North Pole to the South Pole. The Prime Meridian is the meridian passing through the Royal Observatory in Greenwich, England. For this reason the Prime Meridian is sometimes referred to as the **Greenwich Meridian**. This location for the Prime Meridian was adopted at the International Meridian Conference in Washington, D.C., in 1884.

Like latitude, longitude is measured in degrees, minutes, and seconds. For example, the Rand McNally head office is located at 87°43'6" W. The qualifiers E and W indicate whether a location is east or west of the Greenwich Meridian. Longitude ranges from 0° at Greenwich to 180° E or W. The meridian at 180° E is the same as the meridian at 180° W. This meridian, together with the Greenwich Meridian, divides the earth into **eastern and western hemispheres**.

Any circle that divides the earth into equal hemispheres is called a **great circle**. The equator is an example. The shortest distance between any two points on the earth is along a great circle. Other circles, including all other lines of latitude, are called **small circles**. Small circles divide the earth into two unequal pieces.

View of earth centered on 30° N, 30° W

View of earth centered on 30° S, 150° E

The Geographic Grid

The grid of lines of latitude and longitude is known as the **geographic grid**. The following are some important characteristics of the grid.

All lines of longitude are equal in length and meet at the North and South Poles. These lines are called meridians.

All lines of latitude are parallel and equally spaced along meridians. These lines are called parallels.

The length of parallels increases with distance from the poles. For example, the length of the parallel at 60° latitude is one-half the length of the equator.

Meridians get closer together with increasing distance from the equator, and finally converge at the poles.

Parallels and meridians meet at right angles.

Map Scale

To use maps effectively it is important to have a basic understanding of map scale.

Map scale is defined as the ratio of distance on the map to distance on the earth's surface. For example, if a map shows two towns as separated by a distance of 1 inch, and these towns are actually 1 mile apart, then the scale of the map is 1 inch to 1 mile.

The statement "1 inch to 1 mile" is called a **verbal scale**. Verbal scales are simple and intuitive, but a drawback is that they are tied to the specific set of map and real-world units in the numerator and denominator of the ratio. This makes it difficult to compare the scales of different maps.

A more flexible way of expressing scale is as a **representative fraction**. In this case, both the numerator and denominator are converted to the same unit of measurement. For example, since there are 63,360 inches in a mile, the verbal scale "1 inch to 1 mile" can be expressed as the representative fraction 1:63,360. This means that 1 inch on the map represents 63,360 inches on the earth's surface. The advantage of the representative fraction is that it applies to any linear unit of measurement, including inches, feet, miles, meters, and kilometers.

Map scale can also be represented in graphical form. Many maps contain a **graphic scale** (or **bar scale**) showing real-world units such as miles or kilometers. The bar scale is usually subdivided to allow easy calculation of distance on the map.

Map scale has a significant effect on the amount of detail that can be portrayed on a map. This concept is illustrated here using a series of maps of the Washington, D.C., area. (See the figures to the right.) The scales of these maps range from 1:40,000,000 (top map) to 1:4,000,000 (center map) to 1:25,000 (bottom map). The top map has the **smallest scale** of the three maps, and the bottom map has the **largest scale**.

Note that as scale increases, the area of the earth's surface covered by the map decreases. The smallest-scale map covers thousands of square miles, while the largest-scale map covers only a few square miles within the city of Washington. This means that a given feature on the earth's surface will appear larger as map scale increases. On the smallest-scale map, Washington is represented by a small dot. As scale increases the dot becomes an orange shape representing the built-up area of Washington. At the largest scale Washington is so large that only a portion of it fits on the map.

Because small-scale maps cover such a large area, only the largest and most important features can be shown, such as large cities, major rivers and lakes, and international boundaries. In contrast, large-scale maps contain relatively small features, such as city streets, buildings, parks, and monuments.

Small-scale maps depict features in a more simplified manner than large-scale maps. As map scale decreases, the shapes of rivers and other features must be simplified to allow them to be depicted at a highly reduced size. This simplification process is known as **map generalization**.

Maps in *Goode's Atlas of Human Geography* have a wide range of scales. The smallest scales are used for the world thematic map series, where scales range from approximately 1:200,000,000 to 1:75,000,000. Reference map scales range from a minimum of 1:100,000,000 for world maps to a maximum of 1:1,000,000 for city maps. Most reference maps are regional views with a scale of 1:4,000,000.

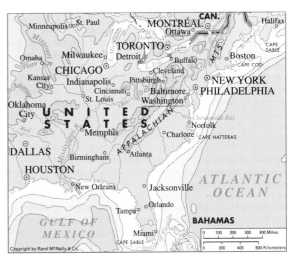

1:40,000,000 scale

1:4,000,000 scale

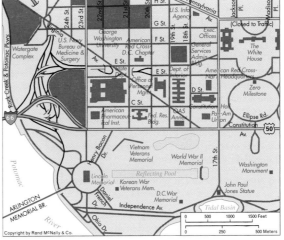

1:25,000 scale

Map Projections

Map projections influence the appearance of features on the map and the ability to interpret geographic phenomena.

A **map projection** is a geometric representation of the earth's surface on a flat or plane surface. Since the earth's surface is curved, a map projection is needed to produce any flat map, whether a page in this atlas or a computer-generated map of driving directions on www.randmcnally.com. Hundreds of projections have been developed since the dawn of mapmaking. A limitation of all projections is that they distort some geometric properties of the earth, such as shape, area, distance, or direction. However, certain properties are preserved on some projections.

If shape is preserved, the projection is called **conformal**. On conformal projections the shapes of features agree with the shapes these features have on the earth. A limitation of conformal projections is that they necessarily distort area, sometimes severely.

Equal-area projections preserve area. On equal area projections the areas of features correspond to their areas on the earth. To achieve this effect, equal-area projections distort shape.

Some projections preserve neither shape nor area, but instead balance shape and area distortion to create an aesthetically-pleasing result. These are often referred to as **compromise** projections.

Distance is preserved on **equidistant** projections, but this can only be achieved selectively, such as along specific meridians or parallels. No projection correctly preserves distance in all directions at all locations. As a result, the stated scale of a map may be accurate for only a limited set of locations. This problem is especially acute for small-scale maps covering large areas.

The projection selected for a particular map depends on the relative importance of different types of distortion, which often depends on the purpose of the map. For example, world maps showing phenomena that vary with area, such as population density or the distribution of agricultural crops, often use an equal-area projection to give an accurate depiction of the importance of each region.

Map projections are created using mathematical procedures. To illustrate the general principles of projections without using mathematics, we can view a projection as the geometric transfer of information from a globe to a flat projection surface, such as a sheet of paper. If we allow the paper to be rolled in different ways, we can derive three basic types of map projections: **cylindrical, conic,** and **azimuthal**. (See the figures to the right.)

For cylindrical projections, the sheet of paper is rolled into a tube and wrapped around the globe so that it is **tangent** (touching) along the equator. Information from the globe is transferred to the tube, and the tube is then unrolled to produce the final flat map.

Conic projections use a cone rather than a cylinder. The figure shows the cone tangent to the earth along a line of latitude with the apex of the cone over the pole. The line of tangency is called the **standard parallel** of the projection.

Azimuthal projections use a flat projection surface that is tangent to the globe at a single point, such as one of the poles.

The figures show the **normal orientation** of each type of surface relative to the globe. The **transverse orientation** is produced when the surface is rotated 90 degrees from normal. For azimuthal projections this orientation is usually called **equatorial** rather than transverse. An **oblique orientation** is created if the projection surface is oriented at an angle between normal and transverse. In general, map distortion increases with distance away from the point or line of tangency. This is why the normal orientations of the cylindrical, conic, and azimuthal projections are often used for mapping equatorial, mid-latitude, and polar regions, respectively.

The projection surface model is a visual tool useful for illustrating how information from the globe can be projected to the map. However, each of the three projection surfaces actually represents scores of individual projections. There are, for example, many projections with the term "cylindrical" in the name, each of which has the same basic rectangular shape, but different spacings of parallels and meridians. The projection surface model does not account for the numerous mathematical details that differentiate one cylindrical, conic, or azimuthal projection from another.

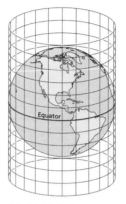

Cylindrical Projection

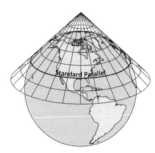

Conic Projection

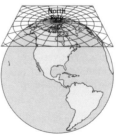
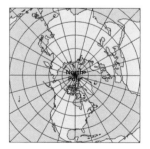

Azimuthal Projection

Map Projections Used in *Goode's Atlas of Human Geography*

Of the hundreds of projections that have been developed, only a fraction are in everyday use. The main projections used in *Goode's Atlas of Human Geography* are described below.

Simple Conic

Type: Conic   **Conformal:** No   **Equal-area:** No

Notes: Shape and area distortion on the Simple Conic projection are relatively low, even though the projection is neither conformal nor equal-area. The origins of the Simple Conic can be traced back nearly two thousand years, with the modern form of the projection dating to the 18th century.

Uses in *Goode's Atlas of Human Geography*: Larger-scale reference maps of North America, Europe, Asia, and other regions.

Lambert Conformal Conic

Type: Conic   **Conformal:** Yes   **Equal-area:** No

Notes: On the Lambert Conformal Conic projection, spacing between parallels increases with distance away from the standard parallel, which allows the property of shape to be preserved. The projection is named after Johann Lambert, an 18th century mathematician who developed some of the most important projections in use today. It became widely used in the United States in the 20th century following its adoption for many statewide mapping programs.

Uses in *Goode's Atlas of Human Geography*: Thematic maps of the United States and Canada, and reference maps of parts of Asia.

Albers Equal-Area Conic

Type: Conic   **Conformal:** No   **Equal-area:** Yes

Notes: On the Albers Equal-Area Conic projection, spacing between parallels decreases with distance away from the standard parallel, which allows the property of area to be preserved. The projection is named after Heinrich Albers, who developed it in 1805. It became widely used in the 20th century, when the United States Coast and Geodetic Survey made it a standard for equal area maps of the United States.

Uses in *Goode's Atlas of Human Geography*: Thematic maps of North America and Asia.

Polyconic

Type: Conic   **Conformal:** No   **Equal-area:** No

Notes: The term polyconic — literally "many-cones" — refers to the fact that this projection is an assemblage of different cones, each tangent at a different line of latitude. In contrast to many other conic projections, parallels are not concentric, and meridians are curved rather than straight. The Polyconic was first proposed by Ferdinand Hassler, who became Head of the United States Survey of the Coast (later renamed the Coast and Geodetic Survey) in 1807. The United States Geological Survey used this projection exclusively for large-scale topographic maps until the mid-20th century.

Uses in *Goode's Atlas of Human Geography*: Reference maps of North America and Asia.

Lambert Azimuthal Equal-Area

Type: Azimuthal   **Conformal:** No   **Equal-area:** Yes

Notes: This projection (another named after Johann Lambert) is useful for mapping large regions, as area is correctly preserved while shape distortion is relatively low. All orientations — polar, equatorial, and oblique — are common.

Uses in *Goode's Atlas of Human Geography*: Thematic and reference maps of North and South America, Asia, Africa, Australia, and polar regions.

Simple Conic Projection

Lambert Conformal Conic Projection

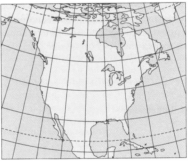

Albers Equal-Area Conic Projection

Polyconic Projection

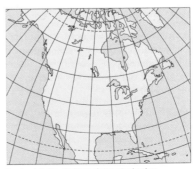

Lambert Azimuthal Equal-Area Projection

Miller Cylindrical

Type: Cylindrical   **Conformal:** No   **Equal-area:** No

Notes: This projection is useful for showing the entire earth in a simple rectangular form. However, polar areas exhibit significant exaggeration of area, a problem common to many cylindrical projections. The projection is named after Osborn Miller, Director of the American Geographical Society, who developed it in 1942 as a compromise projection that is neither conformal nor equal-area.

Uses in *Goode's Atlas of Human Geography*: World climate and time zone maps.

Sinusoidal

Type: Pseudocylindrical   **Conformal:** No   **Equal-area:** Yes

Notes: The straight, evenly spaced parallels on this projection resemble the parallels on cylindrical projections. Unlike cylindrical projections, however, meridians are curved and converge at the poles. This causes significant shape distortion in polar regions. The Sinusoidal is the oldest-known pseudocylindrical projection, dating to the 16th century.

Uses in *Goode's Atlas of Human Geography*: Reference maps of equatorial regions.

Mollweide

Type: Pseudocylindrical   **Conformal:** No   **Equal-area:** Yes

Notes: The Mollweide (or Homolographic) projection resembles the Sinusoidal but has less shape distortion in polar areas due to its elliptical (or oval) form. One of several pseudocylindrical projections developed in the 19th century, it is named after Karl Mollweide, an astronomer and mathematician.

Uses in *Goode's Atlas of Human Geography*: Oceanic reference maps.

Goode's Interrupted Homolosine

Type: Pseudocylindrical   **Conformal:** No   **Equal-area:** Yes

Notes: This projection is a fusion of the Sinusoidal between 40°44'N and S, and the Mollweide between these parallels and the poles. The unique appearance of the projection is due to the introduction of discontinuities in oceanic regions, the goal of which is to reduce distortion for continental landmasses. A condensed version of the projection also exists in which the Atlantic Ocean is compressed in an east-west direction. This modification helps maximize the scale of the map on the page. The Interrupted Homolosine projection is named after J. Paul Goode of the University of Chicago, who developed it in 1923. Goode was an advocate of interrupted projections and, as editor of *Goode's School Atlas*, promoted their use in education.

Uses in *Goode's Atlas of Human Geography*: Small-scale world thematic and reference maps. Both condensed and non-condensed forms are used. An uninterrupted example is used for the Pacific Ocean map.

Robinson

Type: Pseudocylindrical   **Conformal:** No   **Equal-area:** No

Notes: This projection resembles the Mollweide except that polar regions are flattened and stretched out. While it is neither conformal nor equal-area, both shape and area distortion are relatively low. The projection was developed in 1963 by Arthur Robinson of the University of Wisconsin, at the request of Rand McNally.

Uses in *Goode's Atlas of Human Geography*: World maps where the interrupted nature of Goode's Homolosine would be inappropriate, such as the World Oceanic Environments map.

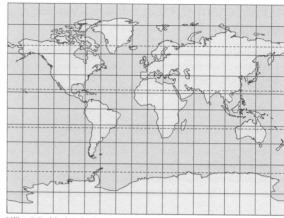

Miller Cylindrical Projection

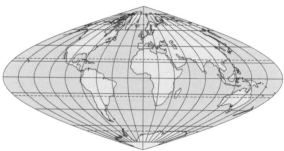

Sinusoidal Projection

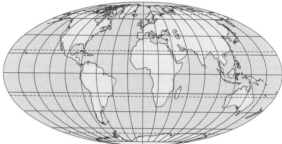

Mollweide Projection

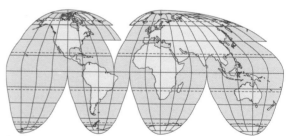

Goode's Interrupted Homolosine Projection

Robinson Projection

Thematic Maps in *Goode's Atlas of Human Geography*

Thematic maps depict a single "theme" such as population density, agricultural productivity, or annual precipitation. The selected theme is presented on a base of locational information, such as coastlines, country boundaries, and major drainage features. The primary purpose of a thematic map is to convey an impression of the overall geographic distribution of the theme. It is usually not the intent of the map to provide exact numerical values. To obtain such information, the graphs and tables accompanying the map should be used.

Goode's Atlas of Human Geography contains many different types of thematic maps. The characteristics of each are summarized below.

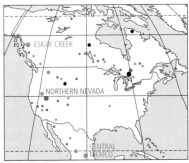

Point symbol map: Detail of Precious Metals

Point Symbol Maps

Point symbol maps are perhaps the simplest type of thematic map. They show features that occur at discrete locations. Examples include earthquakes, nuclear power plants, and minerals-producing areas. The Precious Metals map is an example of a point symbol map showing the locations of areas producing gold, silver, and platinum. A different color is used for each type of metal, while symbol size indicates relative importance.

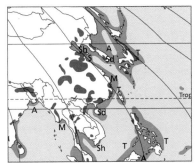

Area symbol map: Detail of Tobacco and Fisheries

Area Symbol Maps

Area symbol maps are useful for delineating regions of interest on the earth's surface. For example, the Tobacco and Fisheries map shows major tobacco-producing regions in one color and important fishing areas in another. On some area symbol maps, different shadings or colors are used to differentiate between major and minor areas.

Dot Maps

Dot maps show a distribution using a pattern of dots, where each dot represents a certain quantity or amount. For example, on the Sugar map, each dot represents 20,000 metric tons of sugar produced. Different dot colors are used to distinguish cane sugar from beet sugar. Dot maps are an effective way of representing the variable density of geographic phenomena over the earth's surface. This type of map is used extensively in *Goode's Atlas of Human Geography* to show the distribution of agricultural commodities.

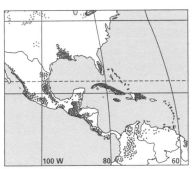

Dot map: Detail of Sugar

Area Class Maps

On area class maps, the earth's surface is divided into areas based on different classes or categories of a particular geographic phenomenon. For example, the Ecoregions map differentiates natural landscape categories, such as Tundra, Savanna, and Prairie. Other examples of area class maps in *Goode's Atlas of Human Geography* include Landforms, Climatic Regions, Natural Vegetation, Soils, Agricultural Areas, Languages and Religions.

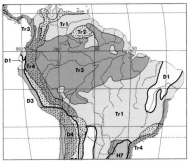

Area class map: Detail of Ecoregions

Isoline Maps

Isoline maps are used to portray quantities that vary smoothly over the surface of the earth. These maps are frequently used for climatic variables such as precipitation and temperature, but a variety of other quantities — from crop yield to population density — can also be treated in this way.

An isoline is a line on the map that joins locations with the same value. For example, the Summer (May to October) Precipitation map contains isolines at 5, 10, 20, and 40 inches. On this map, any 10-inch isoline separates areas that have less than 10 inches of precipitation from areas that have more than 10 inches. Note that the areas between isolines are given different colors to assist in map interpretation.

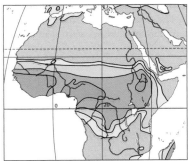

Isoline map: Detail of Precipitation

Proportional Symbol Maps

Proportional symbol maps portray numerical quantities, such as the total population of each state, the total value of agricultural goods produced in different regions, or the amount of hydroelectricity generated in different countries. The symbols on these maps — usually circles —- are drawn such that the size of each is proportional to the value at that location. For example the Exports map shows the value of goods exported by each country in the world, in millions of U.S. dollars.

Proportional symbols are frequently subdivided based on the percentage of individual components making up the total. The Exports map uses wedges of different color to show the percentages of various types of exports, such as manufactured articles and raw materials.

Flow Line Maps

Flow line maps show flows between locations. Usually, the thickness of the flow lines is proportional to flow volume. Flows may be physical commodities like petroleum, or less tangible quantities like information. The flow lines on the Mineral Fuels map represent movement of petroleum measured in billions of U.S. dollars. Note that the locations of flow lines may not represent actual physical routes.

Choropleth Maps

Choropleth maps apply distinctive colors to predefined areas, such as counties or states, to represent different quantities in each area. The quantities shown are usually rates, percentages, or densities. For example, the Birth Rate map shows the annual number of births per one thousand people for each country.

Digital Images

Some maps are actually digital images, analogous to the pictures captured by digital cameras. These maps are created from a very fine grid of cells called **pixels**, each of which is assigned a color that corresponds to a specific value or range of values. The population density maps in this atlas are examples of this type. The effect is much like an isoline map, but the isolines themselves are not shown and the resulting geographic patterns are more subtle and variable. This approach is increasingly being used to map environmental phenomena observable from remote sensing systems.

Cartograms

Cartograms deliberately distort map shapes to achieve specific effects. On **area cartograms**, the size of each area, such as a country, is made proportional to its population. Countries with large populations are therefore drawn larger than countries with smaller populations, regardless of the actual size of these countries on the earth.

The world cartogram series in this atlas depicts each country as a rectangle. This is a departure from cartograms in earlier editions of the atlas, which attempted to preserve some of the salient shape characteristics for each country. The advantage of the rectangle method is that it is easier to compare the area of countries when their shapes are consistent.

The cartogram series incorporates choropleth shading on top of the rectangular cartogram base. In this way map readers can make inferences about the relationship between population and another thematic variable, such as HIV-infection rates.

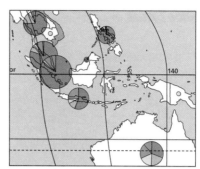
Proportional symbol map: Detail of Exports

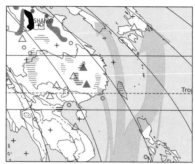
Flow line map: Detail of Mineral Fuels

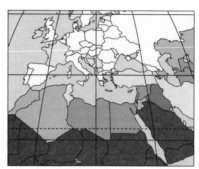
Choropleth map: Detail of Birth Rate

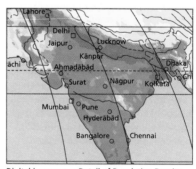
Digital image map: Detail of Population Density

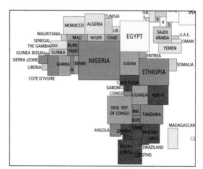
Cartogram: Detail of HIV Infection

GOODE'S

ATLAS OF Human Geography

Map Legend

Political Boundaries

Political maps	Physical maps	
▬▬▬	▬ ▬ ▬	International (Demarcated, Undemarcated, and Administrative)
▬ ▪ ▬	▬ ▪ ▬	Disputed de facto
▬ ▬ ▬	▬ ▬ ▬	Indefinite or Undefined
▬ ▪ ▬	▬ ▪ ▬	Secondary, State, Provincial, etc.

Parks, Indian Reservations

City Limits

Urbanized Areas

Transportation

Political maps	Physical maps	
———	———	Railroads
-------	-------	Railroad Ferries
———		Major Roads
———		Minor Roads
··········		Caravan Routes
✈		Airports

Cultural Features

Dams

Pipelines

▲ Points of Interest

∴ Ruins

Populated Places

◉	1,000,000 and over
◎	250,000 to 1,000,000
⊙	100,000 to 250,000
•	25,000 to 100,000
○	Under 25,000
□	Neighborhoods, Sections of Cities
TŌKYŌ	National Capitals
Boise	Secondary Capitals

Note: On maps at 1:20,000,000 and smaller, symbols do not follow the population classification shown above. Some other maps use a slightly different classification, which is shown in a separate legend in the map margin. On all maps, type size indicates the relative importance of the city.

Land Features

△	Peaks, Spot Heights
⋈	Passes
	Sand
	Contours

Elevation

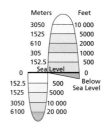

Meters		Feet
3050		10 000
1525		5000
610		2000
305		1000
152.5		500
0	Sea Level	0
152.5		500
1525		5000
3050		10 000
6100		20 000

(Below Sea Level)

Lakes and Reservoirs

Fresh Water

Fresh Water: Intermittent

Salt Water

Salt Water: Intermittent

Other Water Features

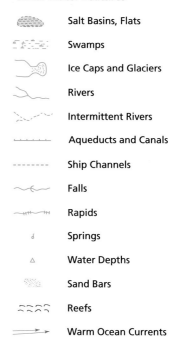

Salt Basins, Flats

Swamps

Ice Caps and Glaciers

Rivers

Intermittent Rivers

Aqueducts and Canals

Ship Channels

Falls

Rapids

Springs

Water Depths

Sand Bars

Reefs

Warm Ocean Currents

Cold Ocean Currents

The legend above shows the symbols used for the political and physical reference maps in *Goode's Atlas of Human Geography*.

To portray relative areas correctly, uniform map scales have been used wherever possible:

        Continents – 1:40,000,000
        Countries and regions – between 1:4,000,000 and 1:20,000,000
        World, polar areas and oceans – between 1:50,000,000
            and 1:100,000,000
        Urbanized areas – 1:1,000,000

Elevations on the maps are shown using a combination of shaded relief and hypsometric tints. Shaded relief (or hill-shading) gives a three-dimensional impression of the landscape, while hypsometric tints show elevation ranges in different colors.

The choice of names for mapped features is complicated by the fact that a variety of languages and alphabets are used throughout the world. A local-names policy is used in *Goode's Atlas of Human Geography* for populated places and local physical features. For some major features, an English form of the name is used with the local name given below in parentheses. Examples include Moscow (Moskva), Vienna (Wien) and Naples (Napoli). In countries where more than one official language is used, names are given in the dominant local language. For large physical features spanning international borders, the conventional English form of the name is used. In cases where a non-Roman alphabet is used, names have been transliterated according to accepted practice.

Selected features are also listed in the Index (pp. 87-92), which includes a pronunciation guide. A list of foreign geographic terms is provided in the Glossary (p. 84).

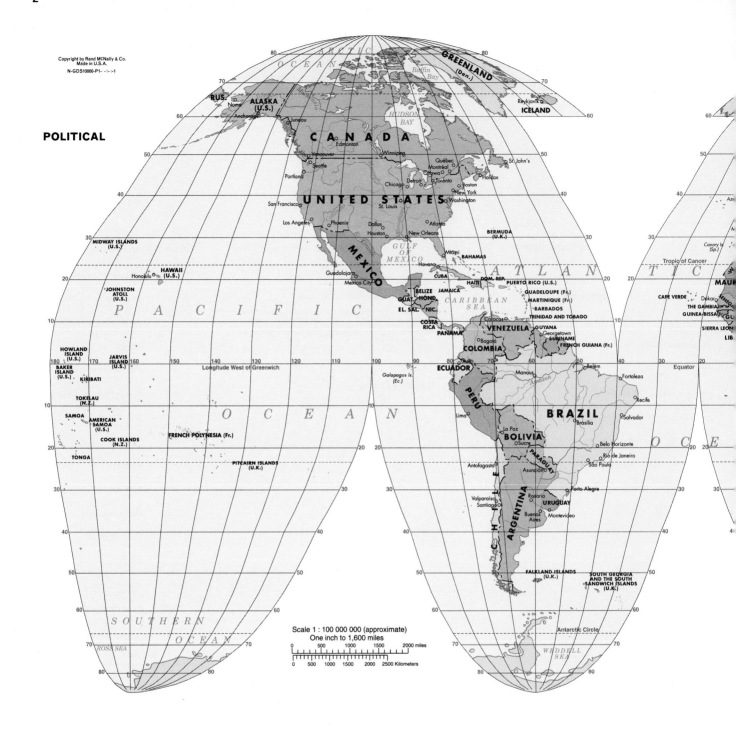

POLITICAL

Scale 1 : 100 000 000 (approximate)
One inch to 1,600 miles

0 500 1000 1500 2000 miles

0 500 1000 1500 2000 2500 Kilometers

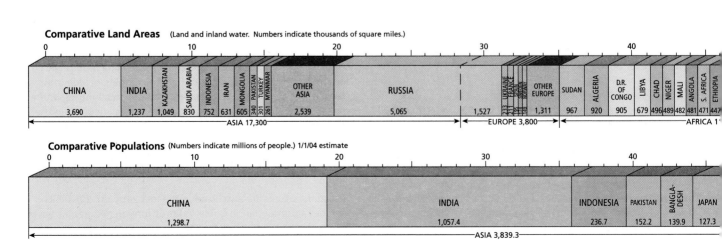

Comparative Land Areas (Land and inland water. Numbers indicate thousands of square miles.)

CHINA	INDIA	KAZAKHSTAN	SAUDI ARABIA	INDONESIA	IRAN	MONGOLIA	PAKISTAN	TURKEY	MYANMAR	OTHER ASIA	RUSSIA	UKRAINE	FRANCE	SPAIN	SWEDEN	NORWAY	OTHER EUROPE	SUDAN	ALGERIA	D.R. OF CONGO	LIBYA	CHAD	NIGER	MALI	ANGOLA	S. AFRICA	ETHIOPIA
3,690	1,237	1,049	830	752	631	605	340	301	261	2,539	5,065	233	211				1,311	967	920	905	679	496	489	482	481	471	447

ASIA 17,300 — EUROPE 3,800 — AFRICA 1

Comparative Populations (Numbers indicate millions of people.) 1/1/04 estimate

CHINA	INDIA	INDONESIA	PAKISTAN	BANGLA-DESH	JAPAN
1,298.7	1,057.4	236.7	152.2	139.9	127.3

ASIA 3,839.3

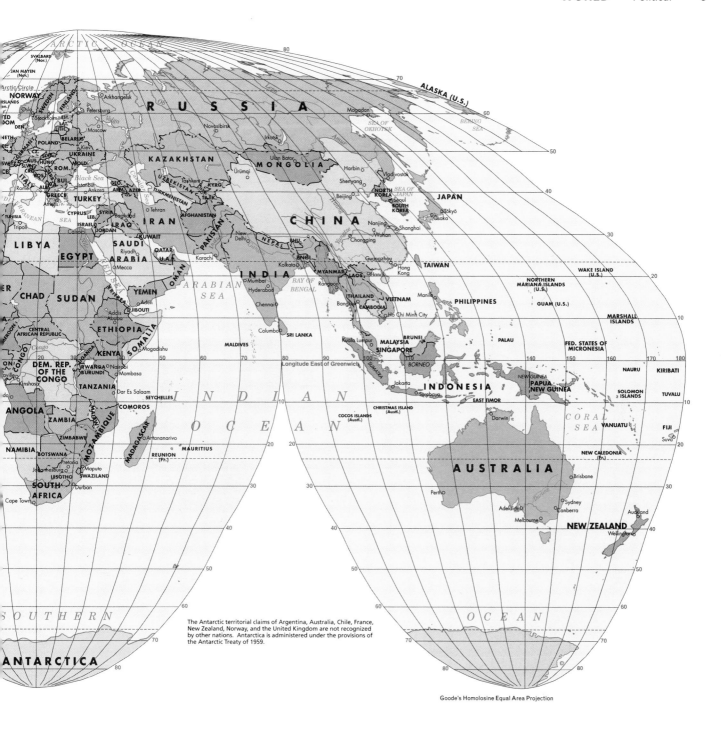

The Antarctic territorial claims of Argentina, Australia, Chile, France,
New Zealand, Norway, and the United Kingdom are not recognized
by other nations. Antarctica is administered under the provisions of
the Antarctic Treaty of 1959.

Goode's Homolosine Equal Area Projection

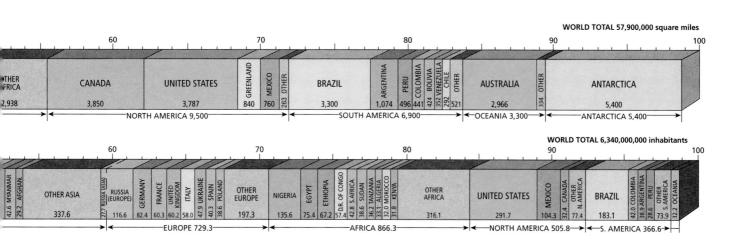

WORLD TOTAL 57,900,000 square miles

	60		70			80						90		100

OTHER AFRICA	CANADA	UNITED STATES	GREENLAND	MEXICO	OTHER	BRAZIL	ARGENTINA	PERU	COLOMBIA	BOLIVIA	VENEZUELA	CHILE	OTHER	AUSTRALIA	OTHER	ANTARCTICA
2,938	3,850	3,787	840	760	263	3,300	1,074	496	441	424	352	292	521	2,966	334	5,400

NORTH AMERICA 9,500 — SOUTH AMERICA 6,900 — OCEANIA 3,300 — ANTARCTICA 5,400

WORLD TOTAL 6,340,000,000 inhabitants

	60		70			80				90		100

MYANMAR	AFGHAN.	OTHER ASIA	RUSSIA (ASIA)	RUSSIA (EUROPE)	GERMANY	FRANCE	UNITED KINGDOM	ITALY	UKRAINE	SPAIN	POLAND	OTHER EUROPE	NIGERIA	EGYPT	ETHIOPIA	D.R. OF CONGO	S. AFRICA	SUDAN	TANZANIA	ALGERIA	MOROCCO	KENYA	OTHER AFRICA	UNITED STATES	MEXICO	CANADA	OTHER N. AMERICA	BRAZIL	COLOMBIA	ARGENTINA	PERU	OTHER S. AMERICA	OCEANIA
42.6	29.2	337.6	27.7	116.6	82.4	60.3	60.2	58.0	47.9	40.3	38.6	197.3	135.6	75.4	67.2	57.4	42.8	38.6	36.2	33.1	32.0	31.8	316.1	291.7	104.3	32.4	77.4	183.1	42.0	38.9	28.6	73.9	32.2

EUROPE 729.3 — AFRICA 866.3 — NORTH AMERICA 505.8 — S. AMERICA 366.6

4

PHYSICAL

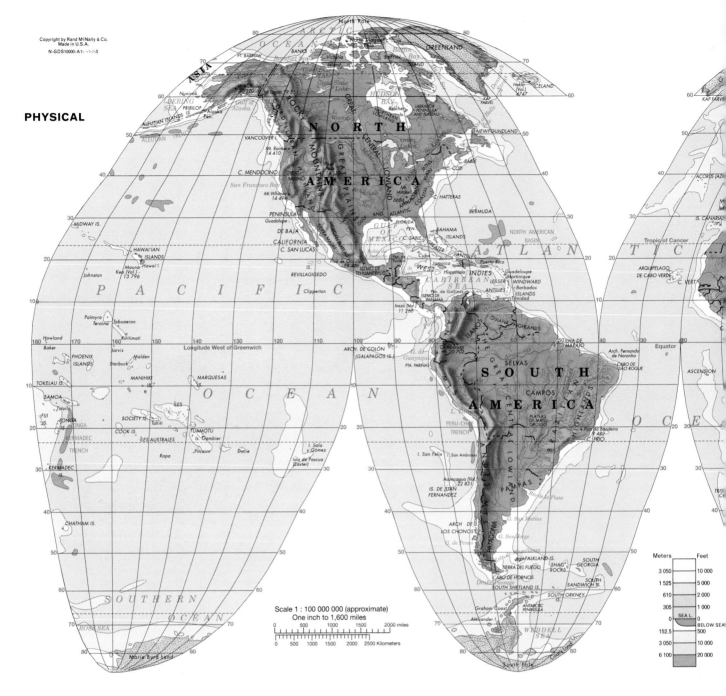

Scale 1 : 100 000 000 (approximate)
One inch to 1,600 miles

Meters	Feet
3 050	10 000
1 525	5 000
610	2 000
305	1 000
0	SEA L. 0
	BELOW SEA
152.5	500
3 050	10 000
6 100	20 000

Land Elevations in Profile

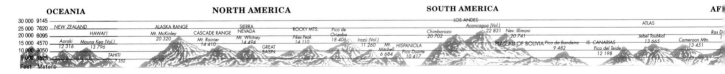

Ocean Depths in Profile

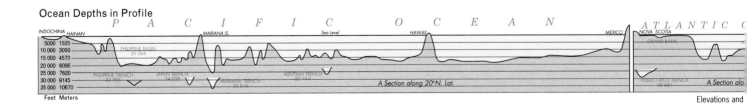

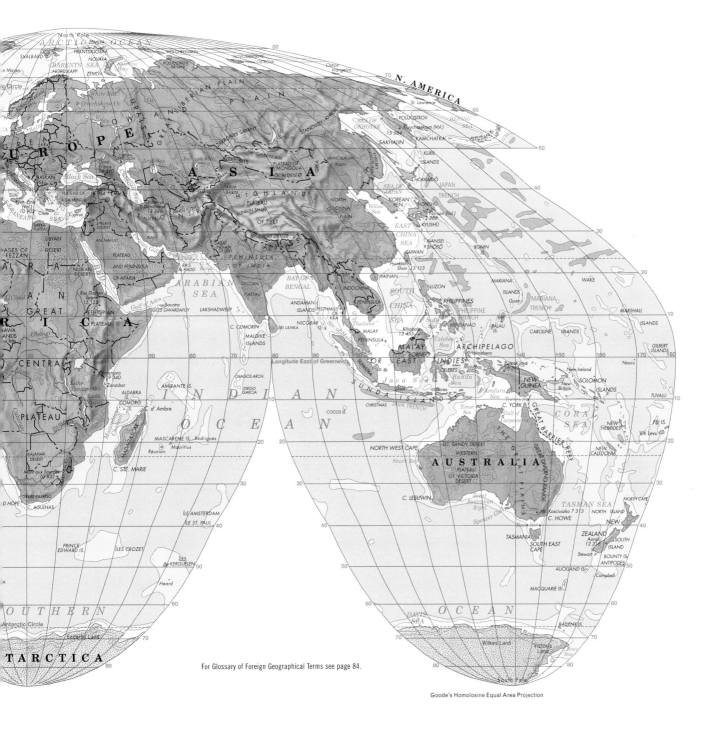

For Glossary of Foreign Geographical Terms see page 84.

Goode's Homolosine Equal Area Projection

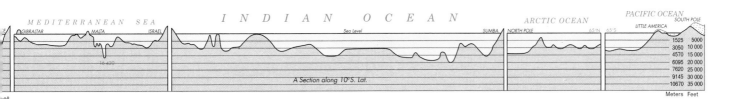

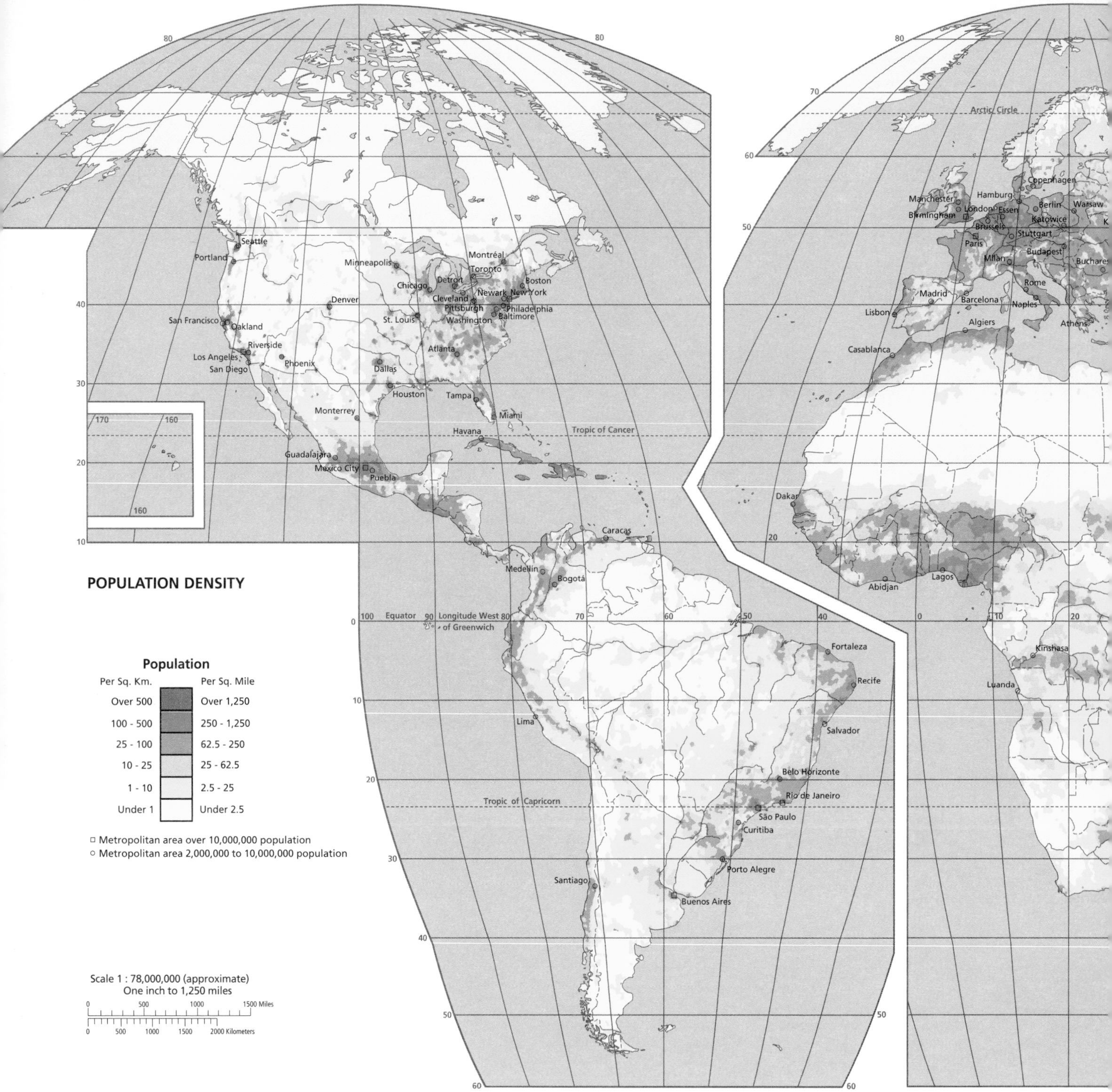
POPULATION DENSITY
Population
Per Sq. Km.
Per Sq. Mile
Over 500
Over 1,250
100 - 500
250 - 1,250
25 - 100
62.5 - 250
10 - 25
25 - 62.5
1 - 10
2.5 - 25
Under 1
Under 2.5
□ Metropolitan area over 10,000,000 population
○ Metropolitan area 2,000,000 to 10,000,000 population
Scale 1 : 78,000,000 (approximate)
One inch to 1,250 miles
0 500 1000 1500 Miles
0 500 1000 1500 2000 Kilometers
Arctic Circle
Tropic of Cancer
Equator
Longitude West of Greenwich
Tropic of Capricorn
Seattle
Portland
San Francisco
Oakland
Riverside
Los Angeles
San Diego
Phoenix
Denver
Minneapolis
Chicago
Detroit
Toronto
Montréal
Boston
New York
Newark
Cleveland
Pittsburgh
Philadelphia
Baltimore
Washington
St. Louis
Atlanta
Dallas
Houston
Tampa
Miami
Monterrey
Havana
Guadalajara
Mexico City
Puebla
Caracas
Medellín
Bogotá
Lima
Fortaleza
Recife
Salvador
Belo Horizonte
Rio de Janeiro
São Paulo
Curitiba
Porto Alegre
Santiago
Buenos Aires
Manchester
Birmingham
London
Hamburg
Essen
Brussels
Paris
Copenhagen
Berlin
Warsaw
Katowice
Stuttgart
Milan
Budapest
Bucharest
Rome
Naples
Athens
Madrid
Barcelona
Lisbon
Algiers
Casablanca
Dakar
Abidjan
Lagos
Kinshasa
Luanda

Largest Countries of the World 1950, 2000, 2050

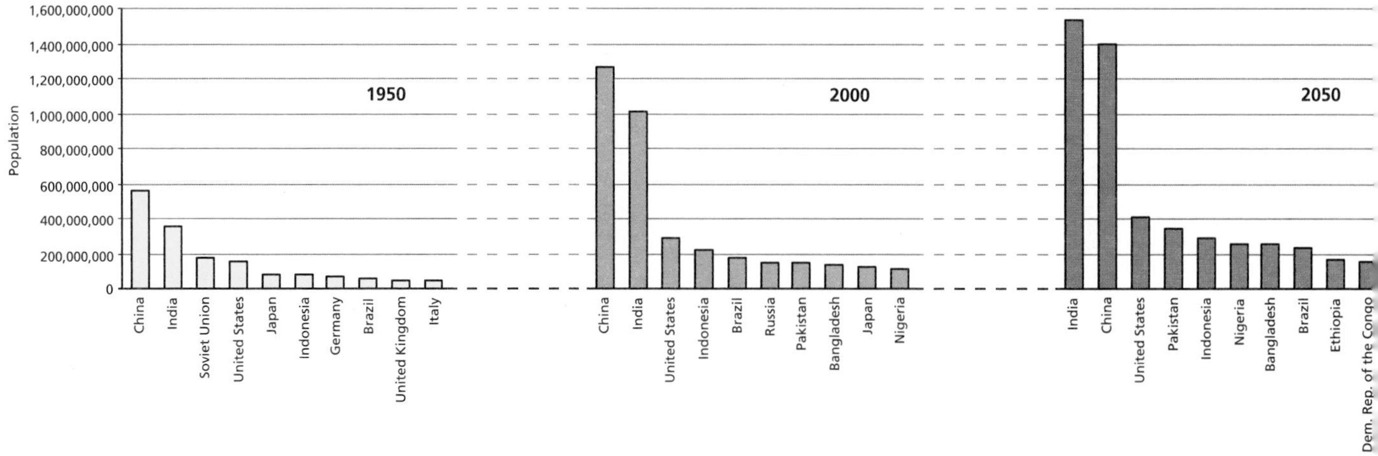
Population
1,600,000,000
1,400,000,000
1,200,000,000
1,000,000,000
800,000,000
600,000,000
400,000,000
200,000,000
0
1950
China
India
Soviet Union
United States
Japan
Indonesia
Germany
Brazil
United Kingdom
Italy
2000
China
India
United States
Indonesia
Brazil
Russia
Pakistan
Bangladesh
Japan
Nigeria
2050
India
China
United States
Pakistan
Indonesia
Nigeria
Bangladesh
Brazil
Ethiopia
Dem. Rep. of the Congo

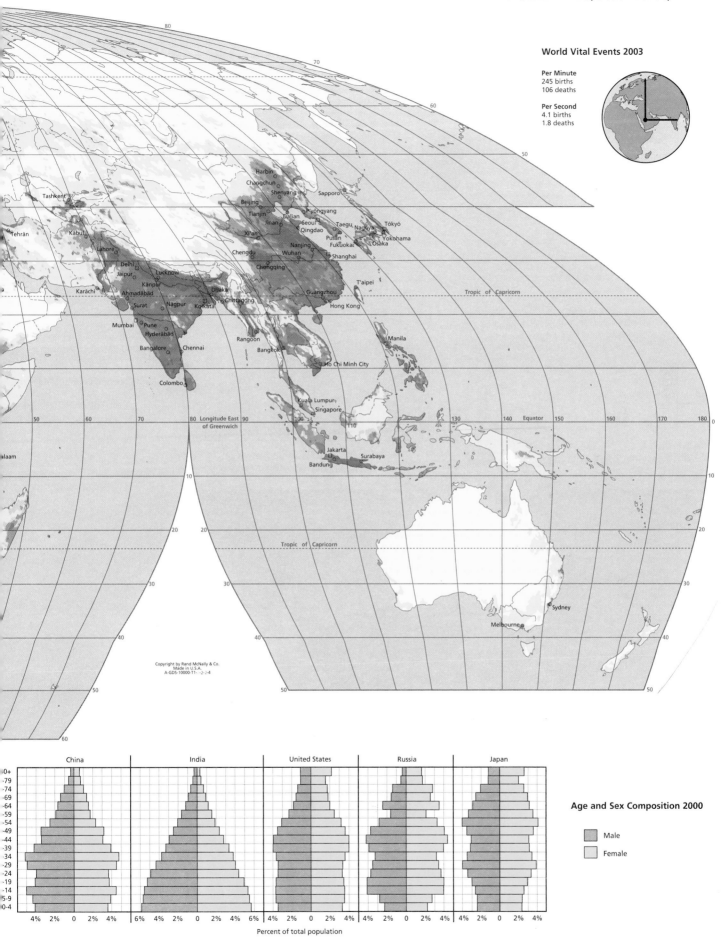

World Vital Events 2003

Per Minute
245 births
106 deaths

Per Second
4.1 births
1.8 deaths

China

India

United States

Russia

Japan

Age and Sex Composition 2000

Male

Female

Percent of total population

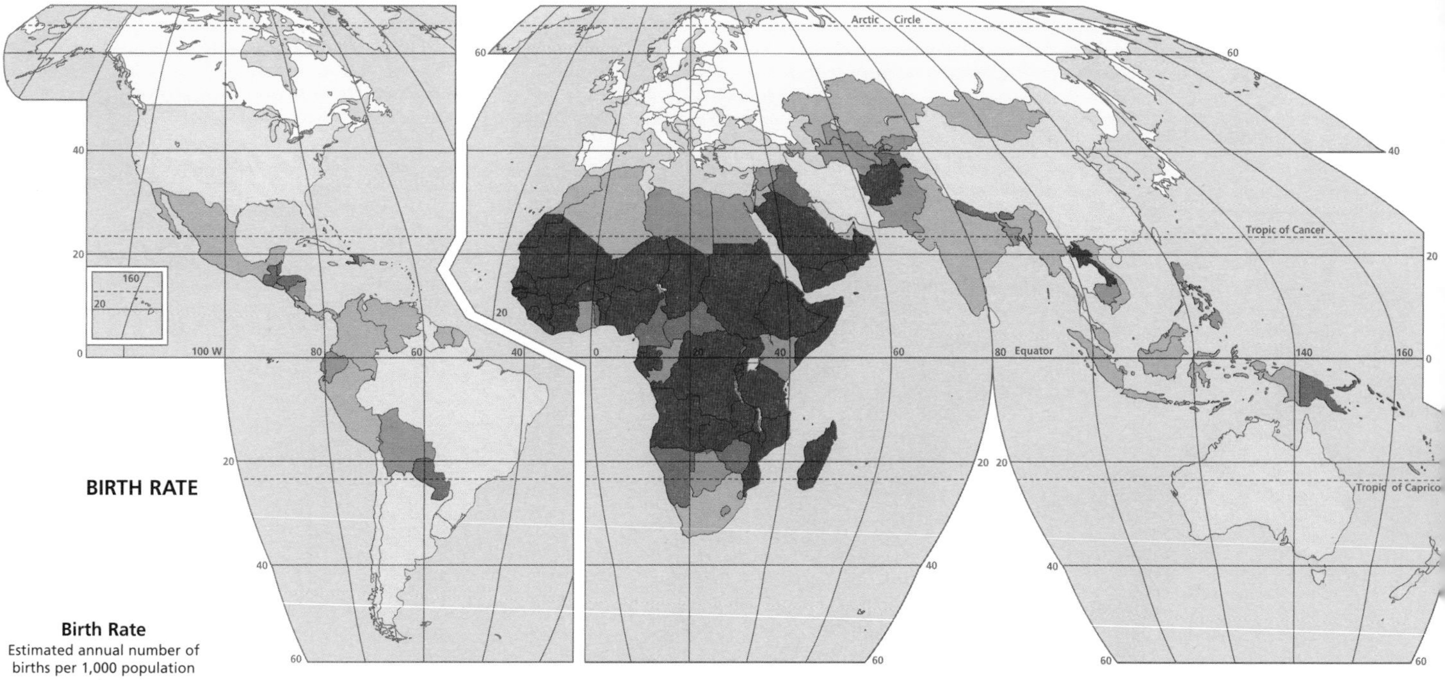

BIRTH RATE

Birth Rate

Estimated annual number of
births per 1,000 population

	Over 36
	30 - 36
	24 - 30
World Avg. 20 →	18 - 24
	12 - 18
	Under 12

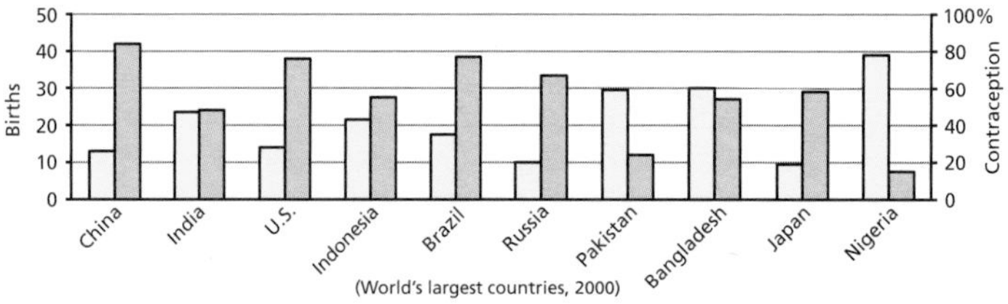

Birth Rate and Contraception Use

	Birth Rate per 1,000 Population
	Percent of Married Women Using Any Type of Contraception

(World's largest countries, 2000)

DEATH RATE

Copyright by Rand McNally & Co.
Made in U.S.A.
A-GDS-10000-U2- -3- -2-4

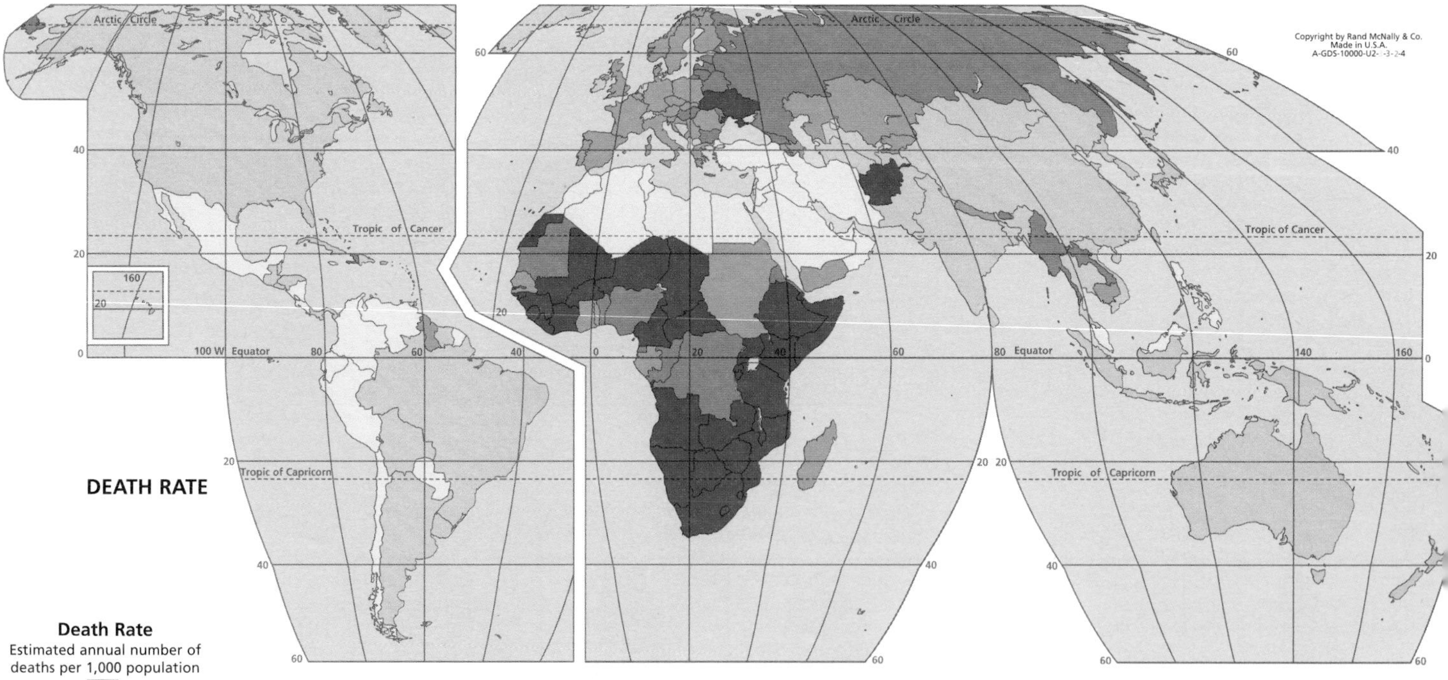

Death Rate

Estimated annual number of
deaths per 1,000 population

	Over 15
	12 - 15
World Avg. 9 →	9 - 12
	6 - 9
	Under 6

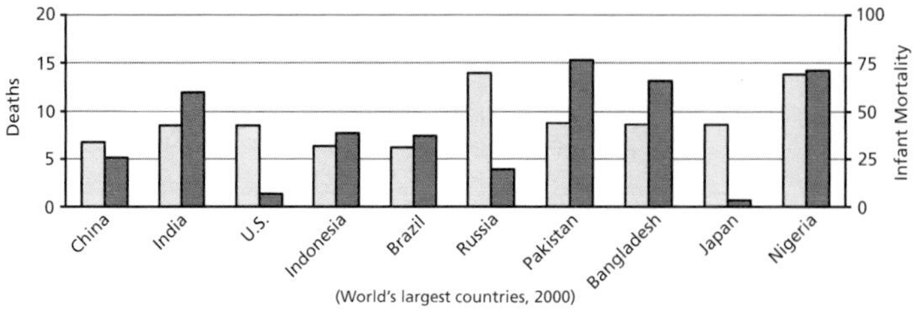

Death Rate and Infant Mortality Rate

	Death Rate per 1,000 Population
	Infant Mortality Rate per 1,000 Live Births

(World's largest countries, 2000)

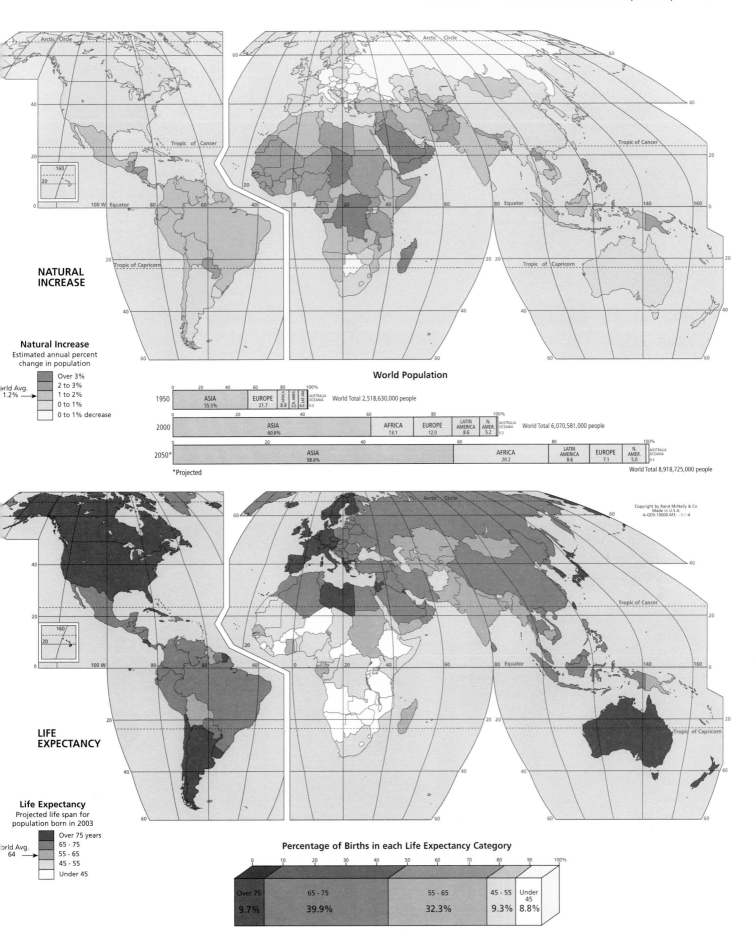

NATURAL INCREASE

Natural Increase
Estimated annual percent change in population

rld Avg.
1.2%

- Over 3%
- 2 to 3%
- 1 to 2%
- 0 to 1%
- 0 to 1% decrease

World Population

1950 | ASIA 55.5% | EUROPE 21.7 | AFRICA 8.8 | N. AMER. 6.6 | LAT. AM. | AUSTRALIA OCEANIA 0.5 | World Total 2,518,630,000 people

2000 | ASIA 60.6% | AFRICA 13.1 | EUROPE 12.0 | LATIN AMERICA 8.6 | N. AMER. 5.2 | AUSTRALIA OCEANIA 0.5 | World Total 6,070,581,000 people

2050* | ASIA 58.6% | AFRICA 20.2 | LATIN AMERICA 8.6 | EUROPE 7.1 | N. AMER. 5.0 | AUSTRALIA OCEANIA 0.5 | World Total 8,918,725,000 people

*Projected

Copyright by Rand McNally & Co.
Made in U.S.A.
A-GDS-10000-M3- -3-1-4

LIFE EXPECTANCY

Life Expectancy
Projected life span for population born in 2003

rld Avg.
64

- Over 75 years
- 65 - 75
- 55 - 65
- 45 - 55
- Under 45

Percentage of Births in each Life Expectancy Category

Over 75	65 - 75	55 - 65	45 - 55	Under 45
9.7%	39.9%	32.3%	9.3%	8.8%

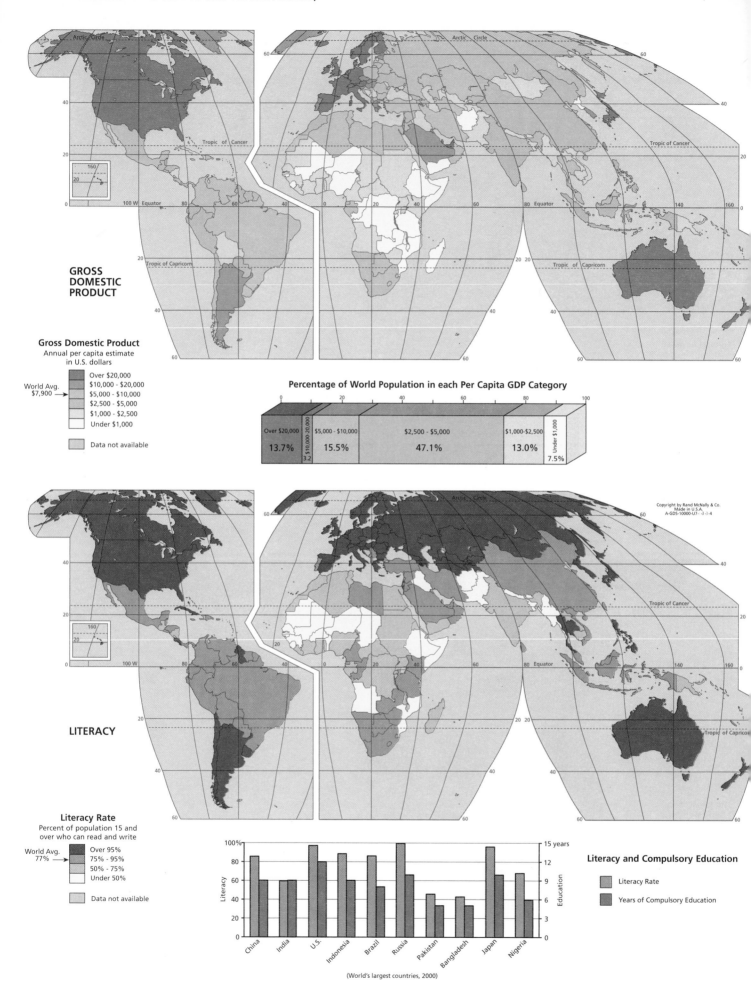

GROSS DOMESTIC PRODUCT

Gross Domestic Product
Annual per capita estimate in U.S. dollars

World Avg. $7,900 →

- Over $20,000
- $10,000 - $20,000
- $5,000 - $10,000
- $2,500 - $5,000
- $1,000 - $2,500
- Under $1,000

Data not available

Percentage of World Population in each Per Capita GDP Category

Over $20,000	$10,000-20,000	$5,000 - $10,000	$2,500 - $5,000	$1,000-$2,500	Under $1,000
13.7%	3.2	15.5%	47.1%	13.0%	7.5%

LITERACY

Literacy Rate
Percent of population 15 and over who can read and write

World Avg. 77% →

- Over 95%
- 75% - 95%
- 50% - 75%
- Under 50%

Data not available

Literacy and Compulsory Education

- Literacy Rate
- Years of Compulsory Education

(World's largest countries, 2000)

Copyright by Rand McNally & Co.
Made in U.S.A.
A-GDS-10000-U7- -3-3-4

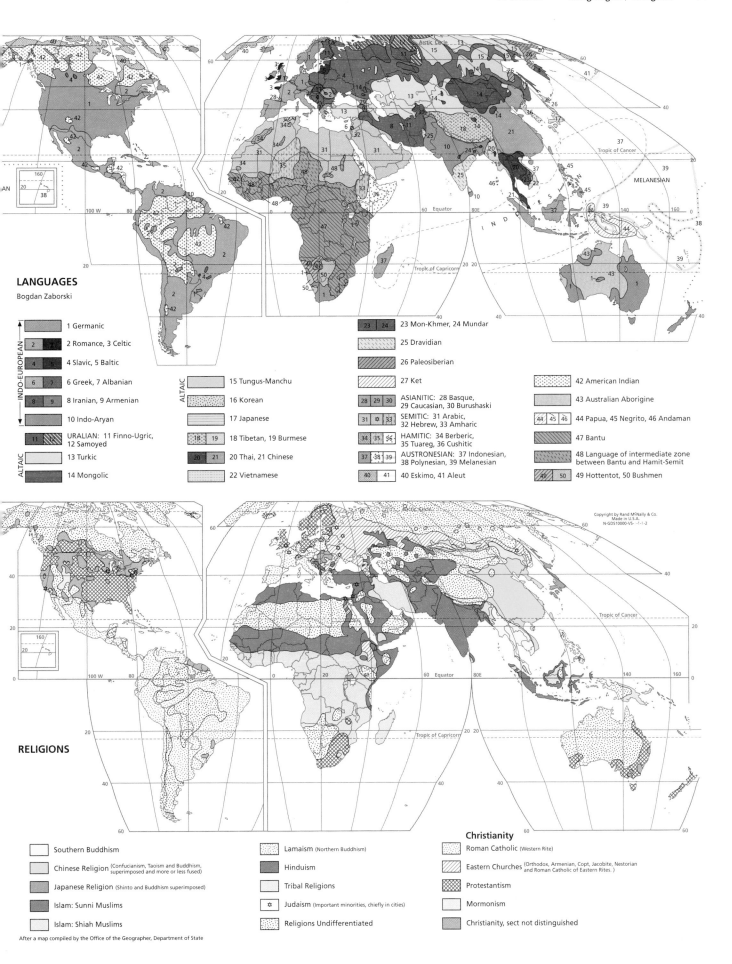

LANGUAGES

Bogdan Zaborski

INDO-EUROPEAN		
	1 Germanic	
	2 Romance, 3 Celtic	
	4 Slavic, 5 Baltic	
	6 Greek, 7 Albanian	
	8 Iranian, 9 Armenian	
	10 Indo-Aryan	
	URALIAN: 11 Finno-Ugric, 12 Samoyed	
ALTAIC	13 Turkic	
	14 Mongolic	

ALTAIC	
	15 Tungus-Manchu
	16 Korean
	17 Japanese
	18 Tibetan, 19 Burmese
	20 Thai, 21 Chinese
	22 Vietnamese

23 24	23 Mon-Khmer, 24 Mundar
	25 Dravidian
	26 Paleosiberian
	27 Ket
28 29 30	ASIANITIC: 28 Basque, 29 Caucasian, 30 Burushaski
31 ✿ 33	SEMITIC: 31 Arabic, 32 Hebrew, 33 Amharic
34 35 36	HAMITIC: 34 Berberic, 35 Tuareg, 36 Cushitic
37 38 39	AUSTRONESIAN: 37 Indonesian, 38 Polynesian, 39 Melanesian
40 41	40 Eskimo, 41 Aleut

	42 American Indian
	43 Australian Aborigine
44 45 46	44 Papua, 45 Negrito, 46 Andaman
	47 Bantu
	48 Language of intermediate zone between Bantu and Hamit-Semit
49 50	49 Hottentot, 50 Bushmen

RELIGIONS

	Southern Buddhism
	Chinese Religion (Confucianism, Taoism and Buddhism, superimposed and more or less fused)
	Japanese Religion (Shinto and Buddhism superimposed)
	Islam: Sunni Muslims
	Islam: Shiah Muslims

	Lamaism (Northern Buddhism)
	Hinduism
	Tribal Religions
✿	Judaism (Important minorities, chiefly in cities)
	Religions Undifferentiated

Christianity

	Roman Catholic (Western Rite)
	Eastern Churches (Orthodox, Armenian, Copt, Jacobite, Nestorian and Roman Catholic of Eastern Rites.)
	Protestantism
	Mormonism
	Christianity, sect not distinguished

After a map compiled by the Office of the Geographer, Department of State

Copyright by Rand McNally & Co.
Made in U.S.A.
N-GD510000-VS- -1-1-2

URBANIZED POPULATION

NORWAY FINLAND
IRELAND UNITED KINGDOM DENMARK SWEDEN LAT. EST.
LITH. BELARUS
NETH. POLAND UKRAINE RUSSIA MONGOLIA NORTH KOREA JAPAN
BEL. GERMANY CZ. SLVK. 6 ROMANIA KAZAKHSTAN SOUTH KOREA
FRANCE SWITZ. HUNG. GEORGIA AZERBAIJAN KYRGYZSTAN CHINA
SLVN. CRO. SERB. BULG. ARMENIA UZBEKISTAN TAJIKISTAN TAIWAN
PORTUGAL SPAIN ITALY BOS. MAC. TURKEY TURKMEN. AFGHANISTAN
ALBANIA GREECE LEB. SYRIA IRAQ IRAN
CANADA TUNISIA ISR. 3 PAKISTAN NEPAL BHUTAN BANGLADESH LAOS
UNITED STATES MOROCCO ALGERIA 9 4 5 U.A.E. CAMBODIA VIETNAM
SAUDI OMAN INDIA MYANMAR
MEXICO CUBA DOMINICAN PUERTO MAURITANIA MALI NIGER CHAD EGYPT ARABIA YEMEN THAILAND PHILIPPINES
REPUBLIC RICO SENEGAL BURK. ERITREA
HAITI THE GAMBIA FASO SUDAN SOMALIA
JAMAICA GUINEA BISSAU GUINEA NIGERIA
GUATEMALA SIERRA LEONE BENIN ETHIOPIA
HONDURAS TRINIDAD AND LIBERIA GHANA MALAYSIA
EL SALVADOR TOBAGO COTE D'IVOIRE CAMEROON SINGAPORE
NICARAGUA VENEZUELA 2
COSTA RICA GABON UGANDA KENYA INDONESIA PAPUA
PANAMA COLOMBIA CONGO NEW GUINEA
ECUADOR DEM. REP. RW. TANZANIA
PERU OF CONGO BUR. SRI LANKA
BRAZIL MADAGASCAR
BOLIVIA ANGOLA ZAMBIA ZIMB. MALAWI MAURITIUS
PARA. 7 1 MOZ. AUSTRALIA
CHILE URUGUAY SOUTH SWAZILAND
ARGENTINA AFRICA LESOTHO NEW ZEALAND

Percent of Population Living in Urban Areas - 2001

- Over 80%
- 60 – 80%
- 40 – 60%
- 20 – 40%
- Under 20%

Copyright by Rand McNally & Co
Made in U.S.A.

Size of each country is proportional to its population.

☐ = 25,000,000 people

Countries with populations under 1,000,000 are not shown.

1 Botswana 6 Moldova
2 Central African Republic 7 Namibia
3 Gaza Strip 8 Togo
4 Jordan 9 West Bank
5 Kuwait

NUTRITION

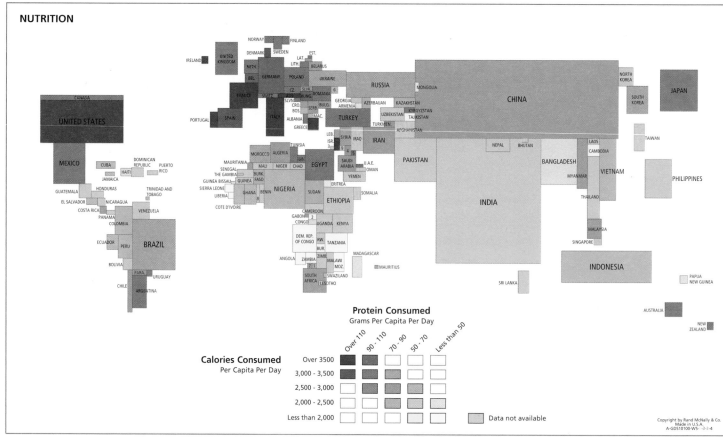

Protein Consumed
Grams Per Capita Per Day

Over 110 | 90 - 110 | 70 - 90 | 50 - 70 | Less than 50

Calories Consumed
Per Capita Per Day

- Over 3500
- 3,000 - 3,500
- 2,500 - 3,000
- 2,000 - 2,500
- Less than 2,000

☐ Data not available

Copyright by Rand McNally & Co.
Made in U.S.A.
A-GDS10100-WS-   -3-)-4

PHYSICIANS

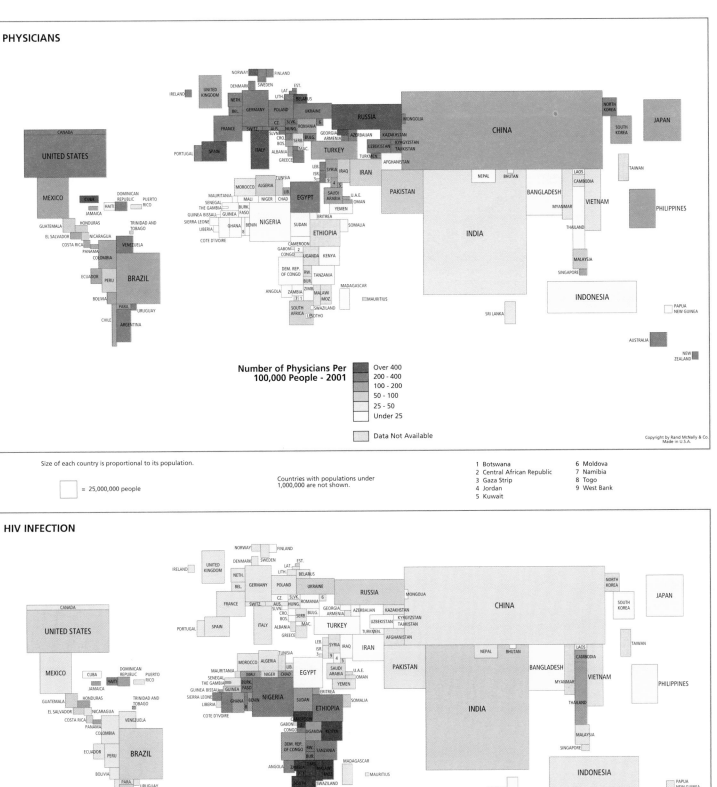

Number of Physicians Per 100,000 People - 2001

- Over 400
- 200 - 400
- 100 - 200
- 50 - 100
- 25 - 50
- Under 25

Data Not Available

Copyright by Rand McNally & Co.
Made in U.S.A.

Size of each country is proportional to its population.

= 25,000,000 people

Countries with populations under 1,000,000 are not shown.

1 Botswana	6 Moldova
2 Central African Republic	7 Namibia
3 Gaza Strip	8 Togo
4 Jordan	9 West Bank
5 Kuwait	

HIV INFECTION

Percent of Adult Population Diagnosed HIV-Positive

- Over 10%
- 5 - 10%
- 1 - 5%
- 0.5 - 1%
- 0.1 - 0.5%
- Under 0.1%

Data Not Available

Copyright by Rand McNally & Co.
Made in U.S.A.
A-GDS10100-W3--3-/-4

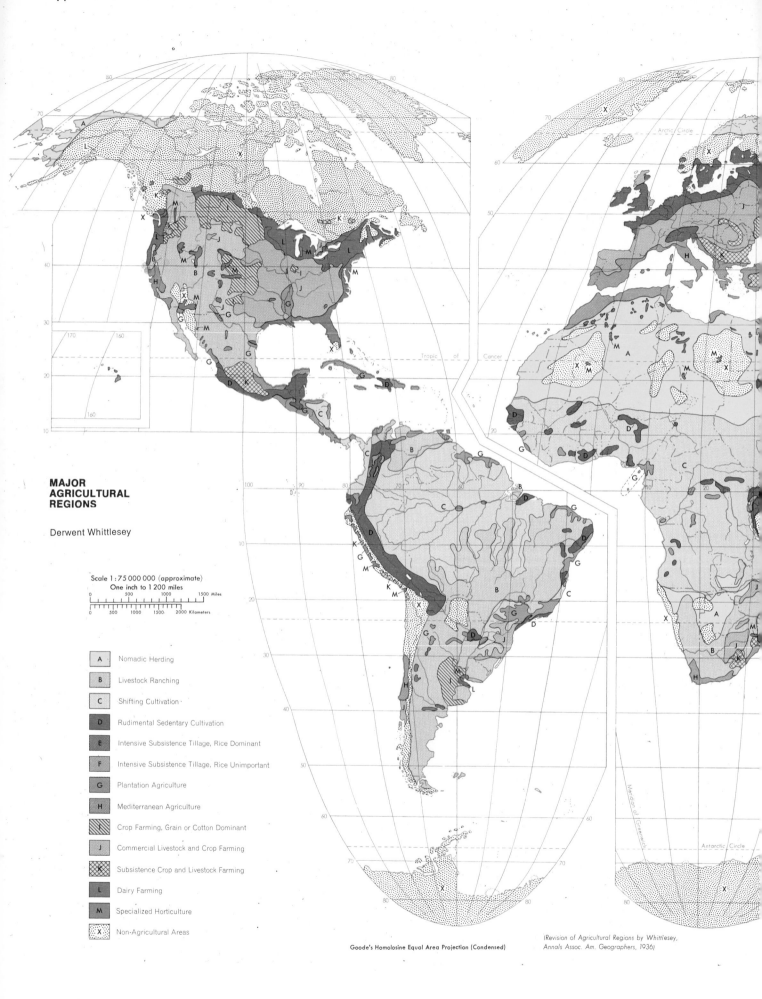

MAJOR AGRICULTURAL REGIONS

Derwent Whittlesey

Scale 1 : 75 000 000 (approximate)
One inch to 1 200 miles

0   500   1000   1500 Miles

0   500   1000   1500   2000 Kilometers

A	Nomadic Herding
B	Livestock Ranching
C	Shifting Cultivation
D	Rudimental Sedentary Cultivation
E	Intensive Subsistence Tillage, Rice Dominant
F	Intensive Subsistence Tillage, Rice Unimportant
G	Plantation Agriculture
H	Mediterranean Agriculture
I	Crop Farming, Grain or Cotton Dominant
J	Commercial Livestock and Crop Farming
K	Subsistence Crop and Livestock Farming
L	Dairy Farming
M	Specialized Horticulture
X	Non-Agricultural Areas

Goode's Homolosine Equal Area Projection (Condensed)

(Revision of Agricultural Regions by Whittlesey,
Annals Assoc. Am. Geographers, 1936)

A-510000-56- -7

Probable Origins of Cultivated Plants

BEET
OLIVE
GRAPE
ONION GARLIC
LETTUCE

APPLE
ALMOND

SOYBEAN

BARLEY
DATE
FIG
FLAX
LENTIL
WHEAT

BUCKWHEAT

APRICOT PEACH
TEA GINGER
RICE

SUGAR BAMBOO
CANE RICE LIME

LEMON

ORANGE
GRAPEFRUIT

BANANA

MILLET
COLA RICE
YAM OKRA
OIL
PALM

SORGHUM

COTTON COFFEE

CLOVE SUGAR
NUTMEG CANE COCONUT

WATERMELON

FORAGE
GRASSES

AVOCADO
CACAO
COMMON BEANS
COTTON
MAIZE
PEPPER
SQUASH
SUNFLOWER
SWEET POTATO
TOBACCO
TOMATO

POTATO
PEANUT
TOMATO

PEANUT
SQUASH
SWEET
POTATO

Hearth Areas

Based on Jack R. Harlan, Crops and Man
(Madison: American Society of Agronomy,
1975) and Erich Isaac, Geography of
Domestication (Prentice Hall, 1970)

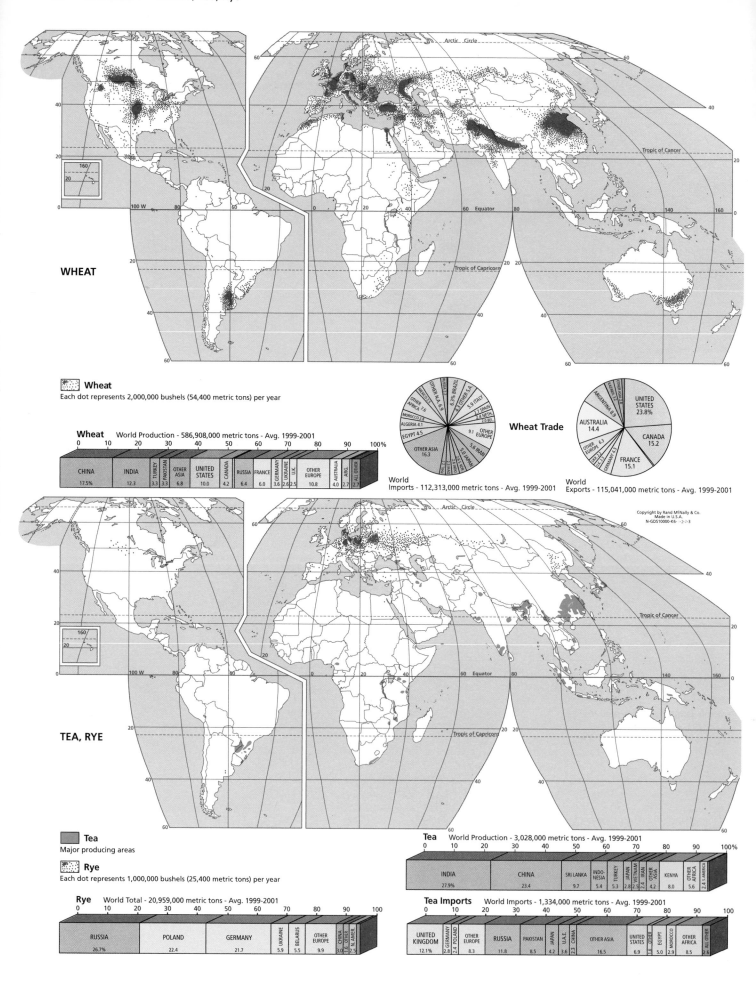

WHEAT

Wheat
Each dot represents 2,000,000 bushels (54,400 metric tons) per year

Wheat World Production - 586,908,000 metric tons - Avg. 1999-2001

0	10	20	30	40	50	60	70	80	90	100%				

CHINA	INDIA	TURKEY	PAKISTAN	OTHER ASIA	UNITED STATES	CANADA	RUSSIA	FRANCE	GERMANY	UKRAINE	U.K.	OTHER EUROPE	AUSTRALIA	ARG.	ALL OTHER
17.5%	12.3	3.3	3.3	6.8	10.0	4.2	6.4	6.0	3.6	2.6	2.5	10.8	4.0	2.7	2.7

Wheat Trade

World
Imports - 112,313,000 metric tons - Avg. 1999-2001

Wheat Trade pie chart: MOROCCO 2.3, OTHER AFRICA 7.6, ALGERIA 4.1, EGYPT 4.5, OTHER ASIA 16.3, 5.0 JAPAN, 5.6 IRAN, OTHER EUROPE 9.1, 2.0 BEL, 1.4 NETH, 3.1 SPAIN, 5.9 ITALY, 5.5 OTHER S.A., 6.3 BRAZIL, 6.3% OTHER N.A.

World
Exports - 115,041,000 metric tons - Avg. 1999-2001

Exports pie chart: UNITED STATES 23.8%, CANADA 15.2, FRANCE 15.1, GERMANY 4.3, U.K. 2.3, OTHER EUROPE 6.3, AUSTRALIA 14.4, ARGENTINA 6.9, OTHER ASIA 1.7

Copyright by Rand McNally & Co.
Made in U.S.A.
N-GDS10000-K6--2-2-3

TEA, RYE

Tea
Major producing areas

Rye
Each dot represents 1,000,000 bushels (25,400 metric tons) per year

Rye World Total - 20,959,000 metric tons - Avg. 1999-2001

0	10	20	30	40	50	60	70	80	90	100

RUSSIA	POLAND	GERMANY	UKRAINE	BELARUS	OTHER EUROPE	CHINA	OTHER	N. AMER.
26.7%	22.4	21.7	5.9	5.5	9.9	3.0	1.6	2.5

Tea World Production - 3,028,000 metric tons - Avg. 1999-2001

0	10	20	30	40	50	60	70	80	90	100%

INDIA	CHINA	SRI LANKA	INDO-NESIA	TURKEY	JAPAN	IRAN	OTHER ASIA	KENYA	OTHER AFRICA	S. AMERICA
27.9%	23.4	9.7	5.4	5.3	2.8	2.4	4.2	8.0	5.6	2.4

Tea Imports World Imports - 1,334,000 metric tons - Avg. 1999-2001

0	10	20	30	40	50	60	70	80	90	100

UNITED KINGDOM	GERMANY	POLAND	OTHER EUROPE	RUSSIA	PAKISTAN	JAPAN	U.A.E.	CHINA	OTHER ASIA	UNITED STATES	EGYPT	MOROCCO	OTHER AFRICA	ALL OTHER
12.1%	2.8	2.4	8.3	11.8	8.5	4.2	3.6	2.3	16.5	6.9	1.6	5.0	2.9	8.5	2.6

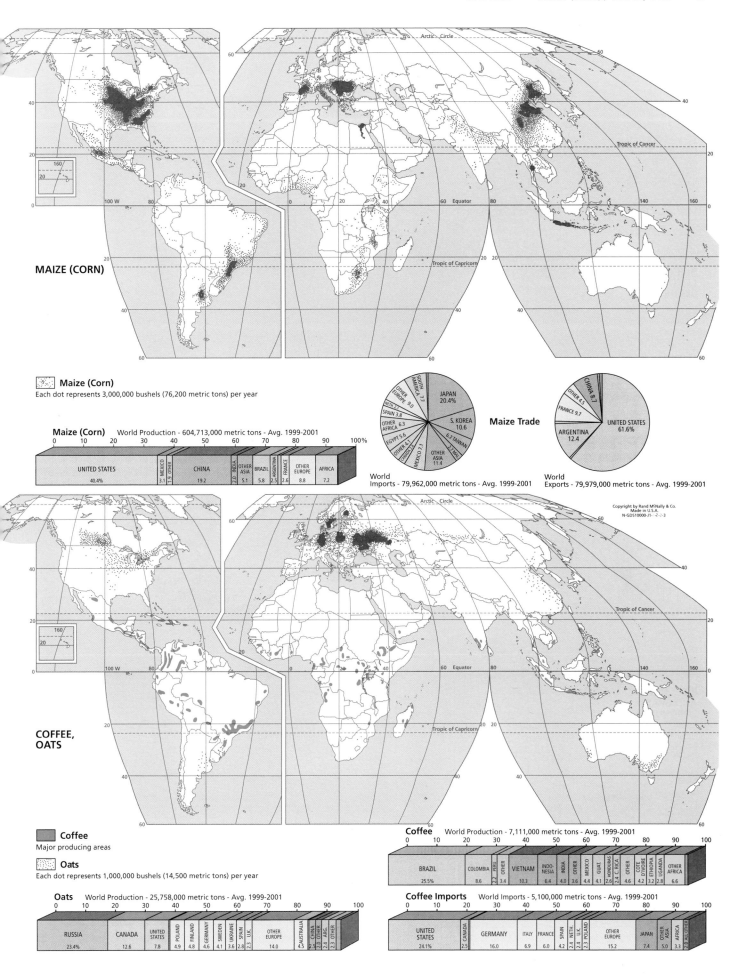

MAIZE (CORN)

Maize (Corn)
Each dot represents 3,000,000 bushels (76,200 metric tons) per year

Maize (Corn) World Production - 604,713,000 metric tons - Avg. 1999-2001

0	10	20	30	40	50	60	70	80	90	100%

UNITED STATES	MEXICO	OTHER	CHINA	INDIA	OTHER ASIA	BRAZIL	ARGENTINA	FRANCE	OTHER EUROPE	AFRICA
40.4%	3.1	1.9	19.2	2.0	5.1	5.8	2.5	2.6	8.8	7.2

Maize Trade

World
Imports - 79,962,000 metric tons - Avg. 1999-2001

JAPAN 20.4%
S. KOREA 10.6
6.2 TAIWAN
2.7 MAL.
OTHER ASIA 11.4
MEXICO 7.1
CANADA 3.1
OTHER 4.1
EGYPT 5.6
OTHER AFRICA 6.3
SPAIN 3.8
NETH. 2.2
OTHER EUROPE 9.0
SOUTH AMERICA 7.7

World
Exports - 79,979,000 metric tons - Avg. 1999-2001

CHINA 8.7
OTHER 4.5
FRANCE 9.7
UNITED STATES 61.6%
ARGENTINA 12.4

COFFEE, OATS

Copyright by Rand McNally & Co.
Made in U.S.A.
N-GD510000-J1- -2- -3

Coffee
Major producing areas

Oats
Each dot represents 1,000,000 bushels (14,500 metric tons) per year

Coffee World Production - 7,111,000 metric tons - Avg. 1999-2001

0	10	20	30	40	50	60	70	80	90	100

BRAZIL	COLOMBIA	PERU	OTHER	VIETNAM	INDO-NESIA	INDIA	OTHER	MEXICO	GUAT.	HONDURAS	C. RICA	CÔTE D'IVOIRE	ETHIOPIA	UGANDA	OTHER AFRICA
25.5%	8.6	2.2	3.4	10.3	6.4	4.0	3.6	4.4	4.1	2.6	2.4	4.6	4.2	3.2	6.6

Oats World Production - 25,758,000 metric tons - Avg. 1999-2001

0	10	20	30	40	50	60	70	80	90	100

RUSSIA	CANADA	UNITED STATES	POLAND	FINLAND	GERMANY	SWEDEN	UKRAINE	SPAIN	U.K.	OTHER EUROPE	AUSTRALIA	CHINA	OTHER	ARG.	OTHER
23.4%	12.6	7.8	4.9	4.8	4.6	4.1	3.6	2.8	2.3	14.0	4.5	2.0	2.4	2.3	

Coffee Imports World Imports - 5,100,000 metric tons - Avg. 1999-2001

0	10	20	30	40	50	60	70	80	90	100

UNITED STATES	CANADA	GERMANY	ITALY	FRANCE	SPAIN	NETH.	U.K.	POLAND	OTHER EUROPE	JAPAN	OTHER ASIA	AFRICA	ALL OTHER
24.1%	2.5	16.0	6.9	6.0	4.2	2.4	2.4	2.3	15.2	7.4	5.0	3.3	2.0

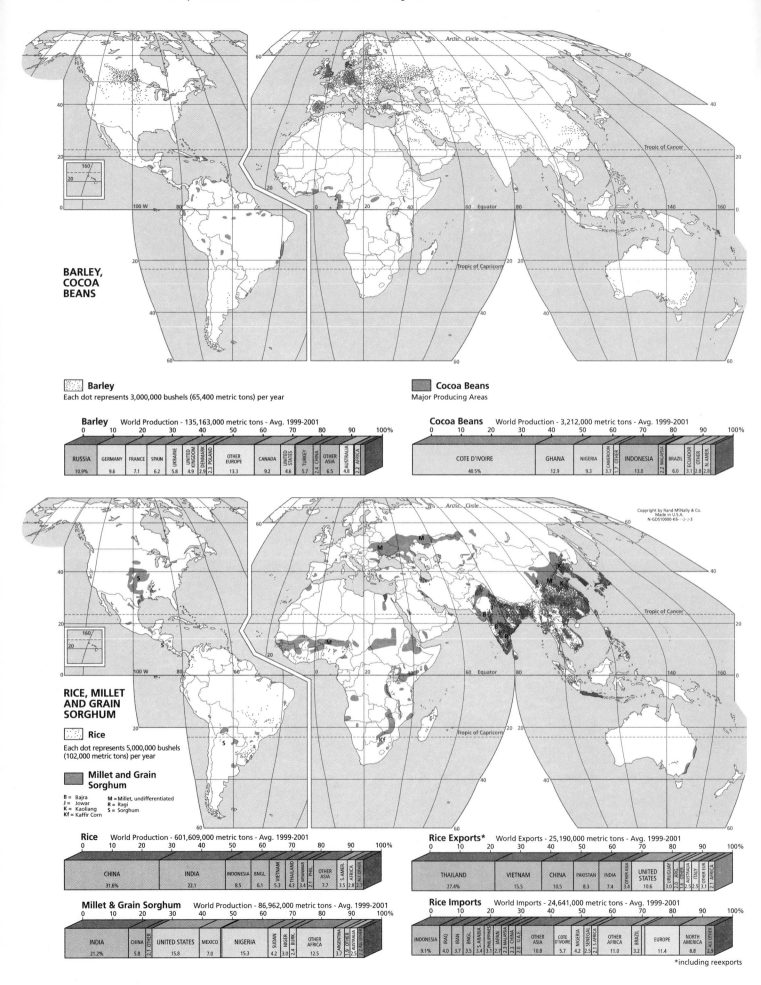

BARLEY, COCOA BEANS

Barley
Each dot represents 3,000,000 bushels (65,400 metric tons) per year

Cocoa Beans
Major Producing Areas

Barley World Production - 135,163,000 metric tons - Avg. 1999-2001

RUSSIA	GERMANY	FRANCE	SPAIN	UKRAINE	UNITED KINGDOM	DENMARK	POLAND	OTHER EUROPE	CANADA	UNITED STATES	TURKEY	CHINA	OTHER ASIA	AUSTRALIA	AFRICA
10.9%	9.6	7.1	6.2	5.8	4.9	2.9	2.3	13.3	9.2	4.6	5.7	2.4	6.5	4.8	2.2

Cocoa Beans World Production - 3,212,000 metric tons - Avg. 1999-2001

COTE D'IVOIRE	GHANA	NIGERIA	CAMEROON	OTHER	INDONESIA	MALAYSIA	BRAZIL	ECUADOR	OTHER / N. AMER.
40.5%	12.9	9.3	3.7	1.7	13.0	2.2	6.0	3.1	2.8 / 2.8

RICE, MILLET AND GRAIN SORGHUM

Rice
Each dot represents 5,000,000 bushels (102,000 metric tons) per year

Millet and Grain Sorghum

B = Bajra
J = Jowar
K = Kaoliang
Kf = Kaffir Corn
M = Millet, undifferentiated
R = Ragi
S = Sorghum

Copyright by Rand McNally & Co.
Made in U.S.A.
N-GDS10000-K6- -2- 2-3

Rice World Production - 601,609,000 metric tons - Avg. 1999-2001

CHINA	INDIA	INDONESIA	BNGL.	VIETNAM	THAILAND	MYANMAR	PHIL.	OTHER ASIA	S. AMER.	AFRICA	ALL OTHER
31.6%	22.1	8.5	6.1	5.3	4.3	3.4	2.1	7.7	3.5	2.8	2.7

Rice Exports* World Exports - 25,190,000 metric tons - Avg. 1999-2001

THAILAND	VIETNAM	CHINA	PAKISTAN	INDIA	OTHER ASIA	UNITED STATES	URUGUAY	ARG.	AUSTRALIA	ITALY / OTHER EUR.	AFRICA
27.4%	15.5	10.5	8.3	7.4	3.4	10.6	3.0	2.0	1.6	2.5 / 2.1	3.1

Millet & Grain Sorghum World Production - 86,962,000 metric tons - Avg. 1999-2001

INDIA	CHINA	OTHER	UNITED STATES	MEXICO	NIGERIA	SUDAN	NIGER	BURK.	OTHER AFRICA	ARGENTINA	OTHER / AUSTRALIA / ALL OTHER
21.2%	5.8	2.1	15.8	7.0	15.3	4.2	3.0	2.4	12.5	3.7	1.9 / 2.5 / 2.2

Rice Imports World Imports - 24,641,000 metric tons - Avg. 1999-2001

INDONESIA	IRAQ	IRAN	BNGL.	S. ARABIA	PHILIPPINES	JAPAN	MALAYSIA	CHINA	U.A.E.	OTHER ASIA	COTE D'IVOIRE	NIGERIA	SENEGAL	S. AFRICA	OTHER AFRICA	BRAZIL	EUROPE	NORTH AMERICA	ALL OTHER
9.1%	4.0	3.7	3.5	3.4	3.1	2.7	2.3	2.3	2.0	10.8	5.7	4.2	2.5	2.1	11.0	3.2	11.4	8.8	2.9

*including reexports

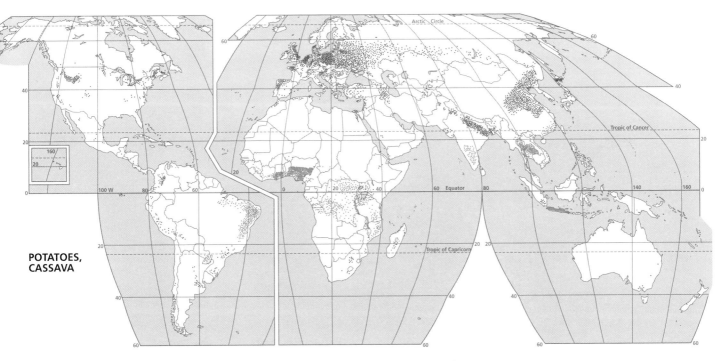

POTATOES, CASSAVA

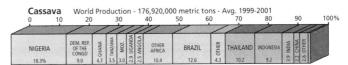

Potatoes
Each dot represents 100,000 metric tons average annual production

Cassava
Each dot represents 100,000 metric tons average annual production

Potatoes — World Total - 312,408,000 metric tons - Avg. 1999-2001

0	10	20	30	40	50	60	70	80	90	100%

CHINA	INDIA	OTHER ASIA	RUSSIA	UNITED STATES	OTHER	POLAND	UKRAINE	GERMANY	BELARUS	NETH.	U.K.	FRANCE	OTHER EUROPE	SOUTH AMERICA	AFRICA
19.9%	7.5	9.5	10.7	6.9	2.1	6.8	5.3	4.0	2.6	2.5	2.2	2.1	8.9	4.4	3.9

Cassava — World Production - 176,920,000 metric tons - Avg. 1999-2001

0	10	20	30	40	50	60	70	80	90	100%

NIGERIA	DEM. REP. OF THE CONGO	GHANA	TANZANIA	MOZ.	UGANDA	ANGOLA	OTHER AFRICA	BRAZIL	OTHER	THAILAND	INDONESIA	INDIA	CHINA	OTHER
18.3%	9.0	4.7	3.5	3.0	2.9	2.5	10.4	12.6	4.3	10.2	9.2	3.9	2.2	2.6

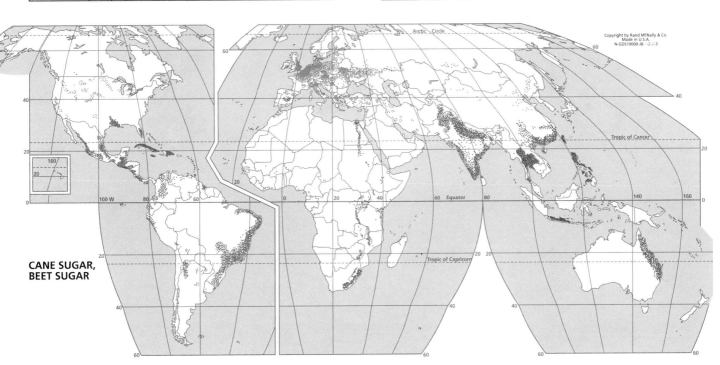

Copyright by Rand McNally & Co.
Made in U.S.A.
N-GDS10000-J8- -2-2-3

CANE SUGAR, BEET SUGAR

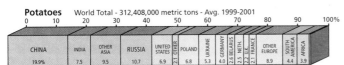
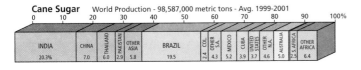

Cane Sugar
Each dot represents 20,000 metric tons average annual production

Beet Sugar
Each dot represents 20,000 metric tons average annual production

Cane Sugar — World Production - 98,587,000 metric tons - Avg. 1999-2001

0	10	20	30	40	50	60	70	80	90	100%

INDIA	CHINA	THAILAND	PAKISTAN	OTHER ASIA	BRAZIL	COL.	OTHER S.A.	MEXICO	CUBA	UNITED STATES	OTHER N.A.	AUSTRALIA	S. AFRICA	OTHER AFRICA
20.3%	7.0	6.0	2.9	5.8	19.5	2.4	4.3	5.2	3.9	3.7	4.6	5.0	2.5	6.4

Beet Sugar — World Production - 35,732,000 metric tons - Avg. 1999-2001

0	10	20	30	40	50	60	70	80	90	100%

GERMANY	FRANCE	POLAND	UKRAINE	ITALY	UNITED KINGDOM	SPAIN	NETH.	BELGIUM	OTHER EUROPE	UNITED STATES	TURKEY	CHINA	OTHER ASIA	RUSSIA	AFRICA	ALL OTHER
12.7%	12.6	5.4	4.9	4.6	4.2	3.2	3.2	2.9	12.7	11.6	6.3	2.6	4.4	4.8	2.2	

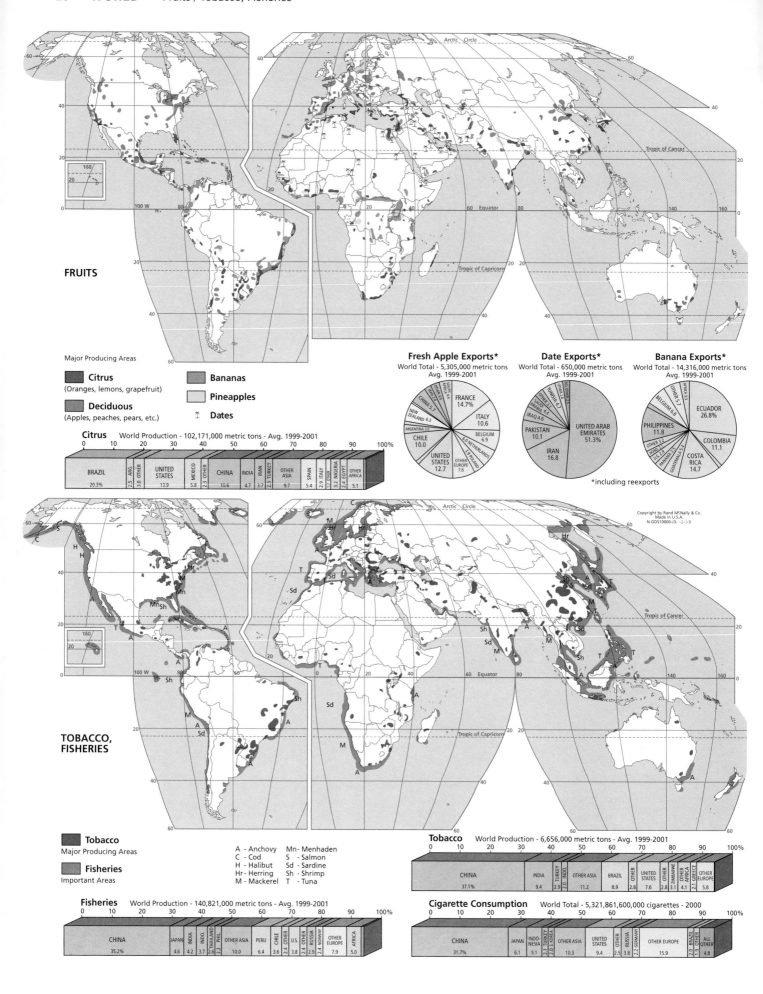

FRUITS

Major Producing Areas

■ **Citrus**
(Oranges, lemons, grapefruit)

■ **Deciduous**
(Apples, peaches, pears, etc.)

■ **Bananas**

■ **Pineapples**

🌴 **Dates**

Citrus World Production - 102,171,000 metric tons - Avg. 1999-2001

BRAZIL	ARG.	OTHER	UNITED STATES	MEXICO	OTHER	CHINA	INDIA	IRAN	TURKEY	OTHER ASIA	SPAIN	ITALY	OTHER	NIGERIA	EGYPT	OTHER AFRICA
20.3%	2.5	3.0	13.9	5.8	2.3	10.6	4.7	3.7	2.9	9.7	5.4	2.9	1.1	3.2	2.4	5.1

Fresh Apple Exports*
World Total - 5,305,000 metric tons
Avg. 1999-2001

SOUTH AFRICA 4.4
CHINA 5.7
NEW ZEALAND 6.3
ARGENTINA 3.0
CHILE 10.0
UNITED STATES 12.7
OTHER EUROPE 7.6
POLAND 3.3
NETHERLANDS 5.0
BELGIUM 6.9
ITALY 10.6
FRANCE 14.7%

Date Exports*
World Total - 650,000 metric tons
Avg. 1999-2001

IRAN 1.3
OTHER 2.1
TUNISIA 4.1
SAUDI ARABIA 4.4
IRAQ 4.6
PAKISTAN 10.1
IRAN 16.8
UNITED ARAB EMIRATES 51.3%

Banana Exports*
World Total - 14,316,000 metric tons
Avg. 1999-2001

AFRICA 3.3
OTHER 5.7
BELGIUM 6.8
PHILIPPINES 11.8
OTHER 3.2
HOND. 2.9
U.S. 2.9
PANAMA 3.1
GUATEMALA 5.3
COSTA RICA 14.7
COLOMBIA 11.1
ECUADOR 26.8%

*including reexports

Copyright by Rand McNally & Co.
Made in U.S.A.
N-GDS10000-/3- -2-2-3

TOBACCO, FISHERIES

■ **Tobacco**
Major Producing Areas

■ **Fisheries**
Important Areas

A - Anchovy Mn - Menhaden
C - Cod   S - Salmon
H - Halibut Sd - Sardine
Hr - Herring Sh - Shrimp
M - Mackerel T - Tuna

Tobacco World Production - 6,656,000 metric tons - Avg. 1999-2001

CHINA	INDIA	TURKEY	INDO.	OTHER ASIA	BRAZIL	OTHER	UNITED STATES	OTHER	ZIMBABWE	OTHER AFRICA	GREECE	OTHER EUROPE
37.1%	9.4	2.9	2.0	11.2	8.9	2.8	7.6	2.8	4.1	2.1		5.8

Fisheries World Production - 140,821,000 metric tons - Avg. 1999-2001

CHINA	JAPAN	INDIA	INDO.	THAILAND	PHIL.	OTHER ASIA	PERU	CHILE	OTHER	U.S.	RUSSIA	NORWAY	OTHER EUROPE	AFRICA
35.2%	4.6	4.2	3.7	2.6	2.2	10.0	6.4	3.6	2.4	3.8	2.4	2.9	7.9	5.0

Cigarette Consumption World Total - 5,321,861,600,000 cigarettes - 2000

CHINA	JAPAN	INDO-NESIA	TURKEY	KOREA	OTHER ASIA	UNITED STATES	OTHER	RUSSIA	GERMANY	OTHER EUROPE	BRAZIL	ALL OTHER
31.7%	6.1	5.1	2.1	2.0	10.3	9.4	2.5	3.8	2.2	15.9	1.9	4.8

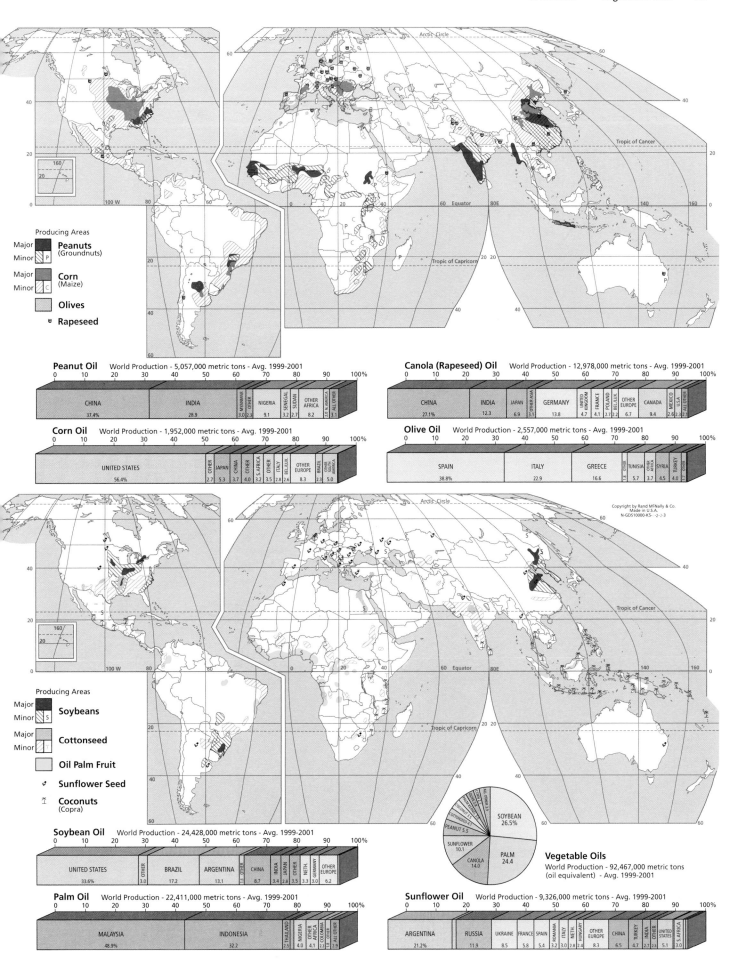

Peanut Oil World Production - 5,057,000 metric tons - Avg. 1999-2001

| CHINA 37.4% | INDIA 28.9 | MYANMAR 3.0 | OTHER 2.3 | NIGERIA 9.1 | SENEGAL 3.2 | SUDAN 2.7 | OTHER AFRICA 8.2 | N. AMERICA 2.0 | ALL OTHER 3.1 |

Corn Oil World Production - 1,952,000 metric tons - Avg. 1999-2001

| UNITED STATES 56.4% | OTHER 2.7 | JAPAN 5.3 | CHINA 3.7 | OTHER 4.0 | S. AFRICA 3.2 | ITALY 3.5 | BELUX 2.6 | OTHER EUROPE 8.3 | BRAZIL 2.3 | OTHER SOUTH AMERICA 5.0 |

Canola (Rapeseed) Oil World Production - 12,978,000 metric tons - Avg. 1999-2001

| CHINA 27.1% | INDIA 12.3 | JAPAN 6.9 | OTHER ASIA 3.1 | GERMANY 13.8 | UNITED KINGDOM 4.7 | FRANCE 4.1 | POLAND 2.7 | BELUX 2.2 | OTHER EUROPE 6.7 | CANADA 9.4 | MEXICO 2.6 | U.S.A. 2.3 | ALL OTHER 2.1 |

Olive Oil World Production - 2,557,000 metric tons - Avg. 1999-2001

| SPAIN 38.8% | ITALY 22.9 | GREECE 16.6 | OTHER 1.8 | TUNISIA 5.7 | OTHER AFRICA 3.7 | SYRIA 4.5 | TURKEY 4.5 | OTHER 1.7 |

Copyright by Rand McNally & Co.
Made in U.S.A.
N-GDS10000-K5- -2-2-3

Soybean Oil World Production - 24,428,000 metric tons - Avg. 1999-2001

| UNITED STATES 33.6% | OTHER 3.0 | BRAZIL 17.2 | ARGENTINA 13.1 | OTHER 1.3 | CHINA 8.7 | INDIA 3.4 | JAPAN 3.5 | NETH. 3.3 | GERMANY 3.0 | OTHER EUROPE 6.2 |

Palm Oil World Production - 22,411,000 metric tons - Avg. 1999-2001

| MALAYSIA 48.9% | INDONESIA 32.2 | THAILAND 2.5 | NIGERIA 4.0 | OTHER AFRICA 4.1 | COLOMBIA 1.3 | OTHER 1.0 | ALL OTHER 2.9 |

Vegetable Oils
World Production - 92,467,000 metric tons
(oil equivalent) - Avg. 1999-2001

Pie chart: SOYBEAN 26.5%, PALM 24.4, CANOLA 14.0, SUNFLOWER 10.1, PEANUT 5.5, COCONUT 4.5, COTTONSEED 4.1, PALM KERNEL 3.3, CORN 2.1, OLIVE, ALL OTHER 3.9

Sunflower Oil World Production - 9,326,000 metric tons - Avg. 1999-2001

| ARGENTINA 21.2% | RUSSIA 11.9 | UKRAINE 8.5 | FRANCE 5.8 | SPAIN 5.4 | ROMANIA 3.2 | ITALY 3.0 | NETH. 2.8 | HUNGARY 2.4 | OTHER EUROPE 8.3 | CHINA 6.5 | TURKEY 4.7 | INDIA 2.7 | OTHER 5.1 | UNITED STATES 5.1 | S. AFRICA 3.0 |

Producing Areas

Peanuts (Groundnuts) — Major / Minor P
Corn (Maize) — Major / Minor C
Olives
Rapeseed

Soybeans — Major / Minor S
Cottonseed — Major / Minor
Oil Palm Fruit
Sunflower Seed
Coconuts (Copra)

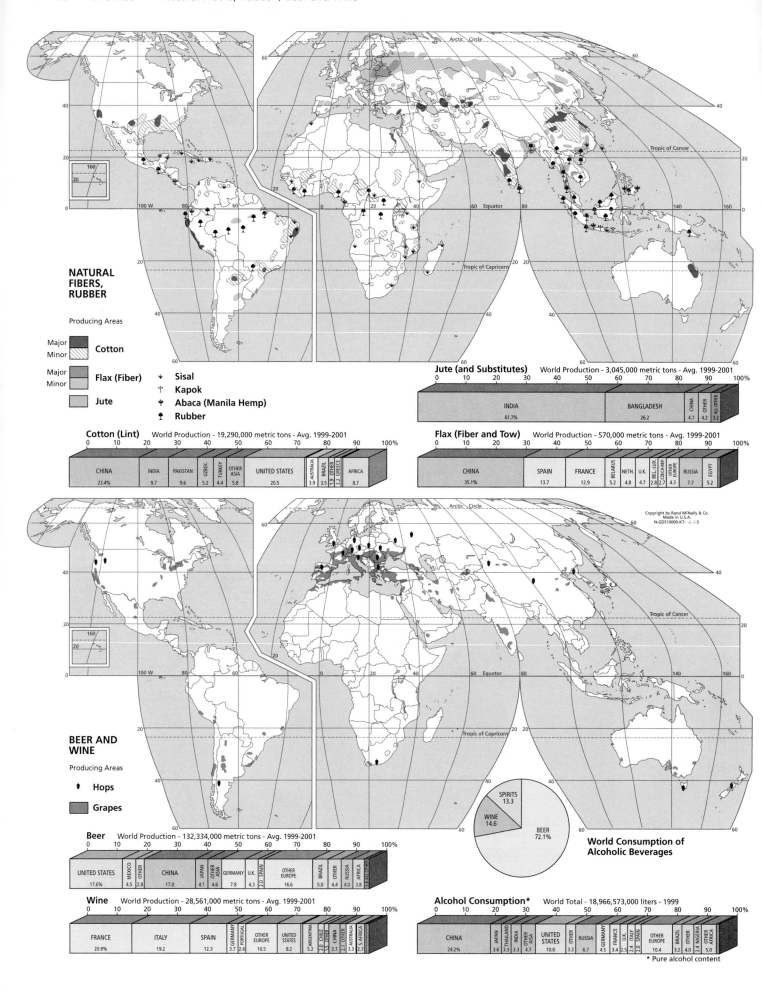

NATURAL FIBERS, RUBBER

Producing Areas

Major / Minor	**Cotton**
Major / Minor	**Flax (Fiber)**
	Jute

↓ Sisal
⸸ Kapok
♣ Abaca (Manila Hemp)
♠ Rubber

Jute (and Substitutes)
World Production - 3,045,000 metric tons - Avg. 1999-2001

INDIA	BANGLADESH	CHINA	OTHER	ALL OTHER
61.7%	26.2	4.7	4.2	3.2

Cotton (Lint)
World Production - 19,290,000 metric tons - Avg. 1999-2001

CHINA	INDIA	PAKISTAN	UZBEK.	TURKEY	OTHER ASIA	UNITED STATES	AUSTRALIA	BRAZIL	OTHER	GREECE	AFRICA
23.4%	9.7	9.6	5.2	4.4	5.8	20.5	3.9	3.5	1.9	2.2	8.7

Flax (Fiber and Tow)
World Production - 570,000 metric tons - Avg. 1999-2001

CHINA	SPAIN	FRANCE	BELARUS	NETH.	U.K.	BEL.-LUX.	CZECH REP.	OTHER EUROPE	RUSSIA	EGYPT
35.1%	13.7	12.9	5.2	4.8	4.7	2.8	2.7	4.3	7.7	5.2

Copyright by Rand McNally & Co.
Made in U.S.A.
N-GDS10000-K7- -2-2-3

BEER AND WINE

Producing Areas

● Hops

▬ Grapes

World Consumption of Alcoholic Beverages

SPIRITS 13.3
WINE 14.6
BEER 72.1%

Beer
World Production - 132,334,000 metric tons - Avg. 1999-2001

UNITED STATES	MEXICO	OTHER	CHINA	JAPAN	OTHER ASIA	GERMANY	U.K.	SPAIN	OTHER EUROPE	BRAZIL	OTHER	RUSSIA	AFRICA	ALL OTHER
17.6%	4.5	2.8	17.0	4.1	4.6	7.8	4.3	2.0	16.6	5.0	4.4	4.0	3.8	1.6

Wine
World Production - 28,561,000 metric tons - Avg. 1999-2001

FRANCE	ITALY	SPAIN	GERMANY	PORTUGAL	OTHER EUROPE	UNITED STATES	ARGENTINA	CHILE	OTHER	CHINA	OTHER	AUSTRALIA	S. AFRICA
20.8%	19.2	12.3	3.7	2.6	10.5	8.2	2.0	1.5	3.7	2.1		3.3	2.7

Alcohol Consumption*
World Total - 18,966,573,000 liters - 1999

CHINA	JAPAN	THAILAND	INDIA	OTHER ASIA	UNITED STATES	OTHER	RUSSIA	GERMANY	FRANCE	U.K.	ITALY	SPAIN	OTHER EUROPE	BRAZIL	OTHER	NIGERIA	OTHER AFRICA
24.2%	3.6	3.3	3.3	4.7	10.0	3.2	6.7	4.5	3.4	2.5	2.0	2.1	10.4	3.2	2.4	2.4	5.0

*Pure alcohol content

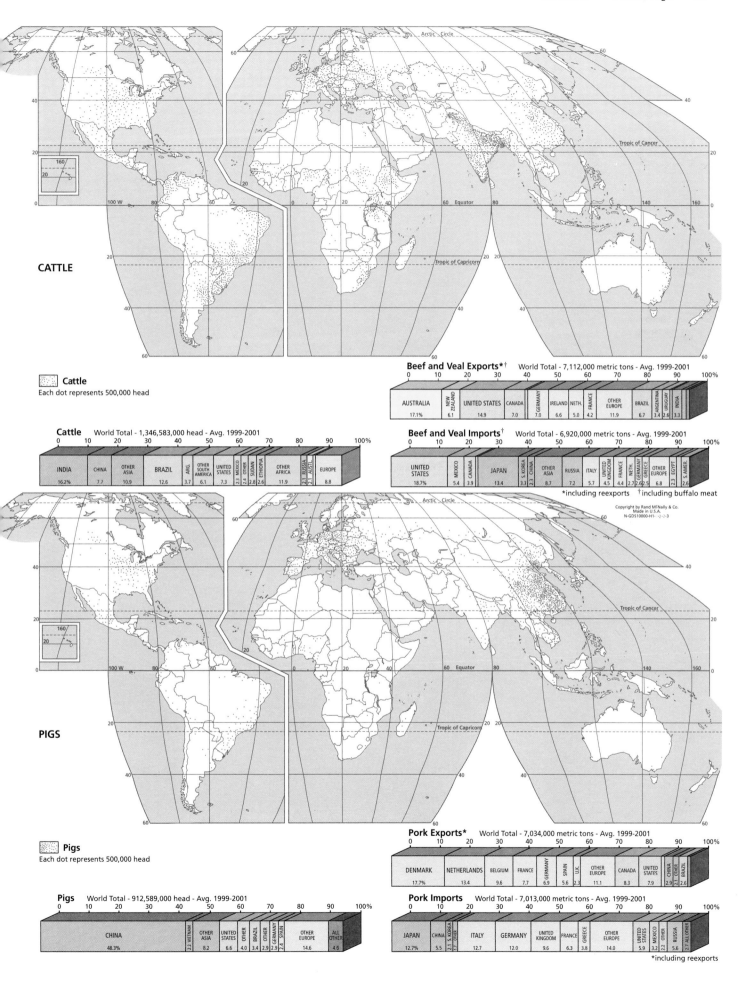

CATTLE

Cattle
Each dot represents 500,000 head

Cattle World Total - 1,346,583,000 head - Avg. 1999-2001

0	10	20	30	40	50	60	70	80	90	100%

INDIA	CHINA	OTHER ASIA	BRAZIL	ARG.	OTHER SOUTH AMERICA	UNITED STATES	MEXICO	OTHER	SUDAN	ETHIOPIA	OTHER AFRICA	RUSSIA	AUSTL.	EUROPE
16.2%	7.7	10.9	12.6	3.7	6.1	7.3	2.3	2.4	2.8	2.6	11.9	2.1	2.1	8.8

Beef and Veal Exports*† World Total - 7,112,000 metric tons - Avg. 1999-2001

0	10	20	30	40	50	60	70	80	90	100%

AUSTRALIA	NEW ZEALAND	UNITED STATES	CANADA	GERMANY	IRELAND	NETH.	FRANCE	OTHER EUROPE	BRAZIL	ARGENTINA	URUGUAY	INDIA
17.1%	6.1	14.9	7.0	7.0	6.6	5.0	4.2	11.9	6.7	3.4	2.6	3.3

Beef and Veal Imports† World Total - 6,920,000 metric tons - Avg. 1999-2001

0	10	20	30	40	50	60	70	80	90	100%

UNITED STATES	MEXICO	CANADA	JAPAN	S. KOREA	CHINA	OTHER ASIA	RUSSIA	ITALY	UNITED KINGDOM	FRANCE	NETH.	GERMANY	GREECE	OTHER EUROPE	EGYPT	S. AMER.
18.7%	5.4	3.9	13.4	3.3	2.3	8.7	7.2	5.7	4.5	4.4	2.7	2.6	2.5	6.8	2.3	2.6

*including reexports †including buffalo meat

Copyright by Rand M℡Nally & Co.
Made in U.S.A.
N-GDS10000-H1- -2-2-3

PIGS

Pigs
Each dot represents 500,000 head

Pigs World Total - 912,589,000 head - Avg. 1999-2001

0	10	20	30	40	50	60	70	80	90	100%

CHINA	VIETNAM	OTHER ASIA	UNITED STATES	OTHER	BRAZIL	OTHER	GERMANY	SPAIN	OTHER EUROPE	ALL OTHER
48.3%	2.3	8.2	6.6	4.0	3.4	2.9	2.4		14.6	4.6

Pork Exports* World Total - 7,034,000 metric tons - Avg. 1999-2001

0	10	20	30	40	50	60	70	80	90	100%

DENMARK	NETHERLANDS	BELGIUM	FRANCE	GERMANY	SPAIN	U.K.	OTHER EUROPE	CANADA	UNITED STATES	CHINA	OTHER	BRAZIL
17.7%	13.4	9.6	7.7	6.9	5.6	2.3	11.1	8.3	7.9	2.9	2.1	2.6

Pork Imports World Total - 7,013,000 metric tons - Avg. 1999-2001

0	10	20	30	40	50	60	70	80	90	100%

JAPAN	CHINA	S. KOREA	OTHER	ITALY	GERMANY	UNITED KINGDOM	FRANCE	GREECE	OTHER EUROPE	UNITED STATES	MEXICO	OTHER	RUSSIA	ALL OTHER
12.7%	5.5	2.1	1.7	12.7	12.0	9.6	6.3	3.8	14.0	5.9	3.2	2.2	5.6	2.7

*including reexports

SHEEP

Arctic Circle
Tropic of Cancer
Equator
Tropic of Capricorn

Sheep
Each dot represents
200,000 head

Sheep
World Total - 1,052,275,000 head - Avg. 1999-2001

0	10	20	30	40	50	60	70	80	90	100%

CHINA	INDIA	IRAN	TURKEY	PAKISTAN	OTHER ASIA	AUSTRALIA	NEW ZEALAND	SUDAN	S. AFRICA	ETHIOPIA	NIGERIA	OTHER AFRICA	UNITED KINGDOM	SPAIN	OTHER EUROPE	SOUTH AMERICA	ALL OTHER
12.4%	5.5	5.1	2.8	2.3	10.3	11.1	4.3	4.4	2.7	2.1	2.0	12.4	3.9	2.3	6.5	7.2	2.7

Wool (Raw)
World Total - 2,313,000 metric tons - Avg. 1999-2001

0	10	20	30	40	50	60	70	80	90	100%

AUSTRALIA	NEW ZEALAND	CHINA	IRAN	INDIA	OTHER ASIA	U.K.	OTHER EUROPE	ARG.	URUGUAY	OTHER	S. AFRICA	SUDAN	OTHER AFRICA	ALL OTHER
29.2%	10.7	12.6	3.2	2.3	12.7	2.7	7.1	2.6	2.5		2.4	2.0	4.6	2.8

Wool Exports (Raw)*
World Total - 666,000 metric tons - Avg. 1999-2001

0	10	20	30	40	50	60	70	80	90	100%

AUSTRALIA	NEW ZEALAND	ARG.	S. AFRICA	EUROPE	ASIA
67.9%	7.4	2.6	2.5	9.4	6.5

Wool Imports (Raw)
World Total - 635,000 metric tons - Avg. 1999-2001

0	10	20	30	40	50	60	70	80	90	100%

CHINA	INDIA	TURKEY	OTHER	ITALY	FRANCE	GERMANY	U.K.	CZECH	SPAIN	OTHER EUROPE	RUSSIA	ALL OTHER
35.4%	4.7	2.9	3.4	16.6	8.2	5.6	5.0	3.1	2.6	6.2	2.8	3.8

*including reexports

POULTRY

Arctic Circle
Tropic of Cancer
Equator
Tropic of Capricorn

Chickens
Each dot represents
1,000,000 chickens

Chickens
World Total - 14,500,527,000 head - Avg. 1999-2001

0	10	20	30	40	50	60	70	80	90	100%

CHINA	INDO-NESIA	INDIA	JAPAN	OTHER ASIA	UNITED STATES	MEXICO	BRAZIL	OTHER S. AMER.	RUSSIA	EUROPE	AFRICA
24.9%	5.4	2.8	2.0	13.8	12.5	3.3	5.8	5.1	2.4	9.9	8.8

Hen Eggs
World Total - 51,406,000 metric tons - Avg. 1999-2001

0	10	20	30	40	50	60	70	80	90	100%

CHINA	JAPAN	INDIA	OTHER ASIA	UNITED STATES	MEXICO	RUSSIA	BRAZIL	FRANCE	OTHER EUROPE	AFRICA
37.7%	5.0	3.5	10.6	9.7	3.4	3.7	2.9	2.0	12.8	4.0

Ducks
World Total - 964,407,000 head - Avg. 1999-2001

0	10	20	30	40	50	60	70	80	90	100%

CHINA	INDIA	VIET-NAM	INDONESIA	THAILAND	OTHER ASIA	FRANCE	UKRAINE	OTHER	ALL OTHER
62.5%	8.3	5.7	3.0	2.7	6.8	2.5	1.9		4.3

Turkeys
World Total - 244,159,000 head - Avg. 1999-2001

0	10	20	30	40	50	60	70	80	90	100%

UNITED STATES	CANADA	OTHER	FRANCE	ITALY	U.K.	GERMANY	PORTUGAL	OTHER EUROPE	BRAZIL	ISRAEL	AFRICA	ALL OTHER
35.5%	2.3	1.7	17.4	10.0	4.0	3.5	2.5	6.7	4.3	2.0	3.6	3.0

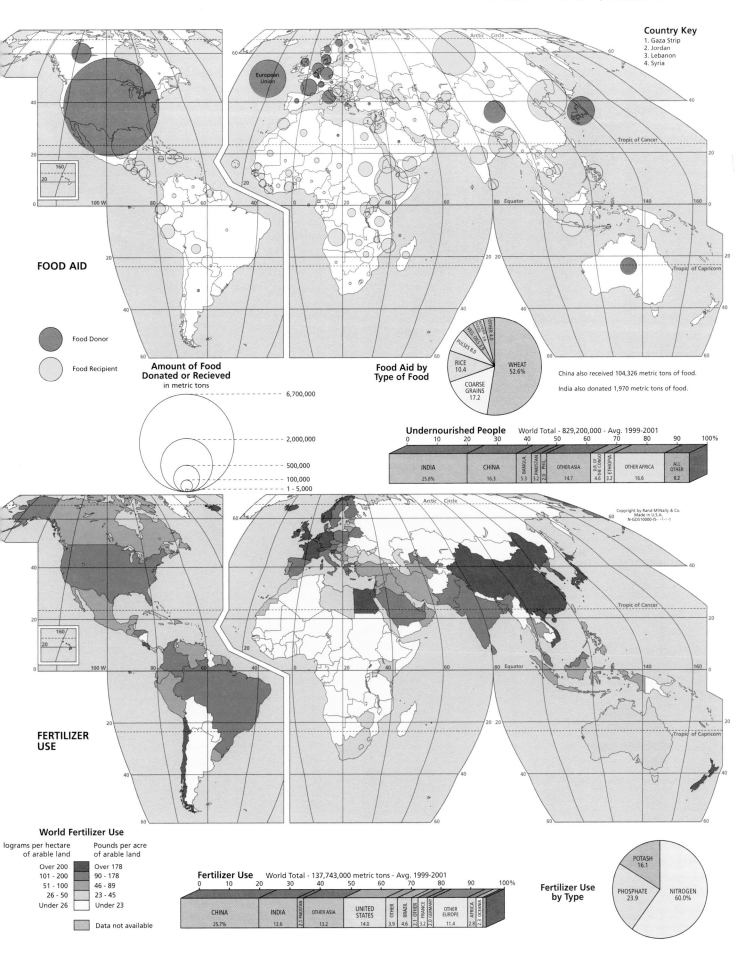

Country Key
1. Gaza Strip
2. Jordan
3. Lebanon
4. Syria

FOOD AID

- Food Donor
- Food Recipient

Amount of Food Donated or Recieved
in metric tons

- 6,700,000
- 2,000,000
- 500,000
- 100,000
- 1 - 5,000

Food Aid by Type of Food

WHEAT 52.6%
RICE 10.4
COARSE GRAINS 17.2
PULSES 8.0
VEG. OILS 3.3
BLENDED 4.6
OTHER 4.0

China also received 104,326 metric tons of food.

India also donated 1,970 metric tons of food.

Undernourished People   World Total - 829,200,000 - Avg. 1999-2001

INDIA	CHINA	BANGLA.	PAKISTAN	PHIL.	OTHER ASIA	D.R. OF THE CONGO	ETHIOPIA	OTHER AFRICA	ALL OTHER
25.8%	16.3	5.3	3.2	2.0	14.7	4.6	3.2	16.6	8.2

Copyright by Rand McNally & Co.
Made in U.S.A.
N-GDS10000-I5- -1-1-1

FERTILIZER USE

World Fertilizer Use

kilograms per hectare of arable land	Pounds per acre of arable land
Over 200	Over 178
101 - 200	90 - 178
51 - 100	46 - 89
26 - 50	23 - 45
Under 26	Under 23

Data not available

Fertilizer Use   World Total - 137,743,000 metric tons - Avg. 1999-2001

CHINA	INDIA	PAKISTAN	OTHER ASIA	UNITED STATES	OTHER	BRAZIL	OTHER	FRANCE	GERMANY	OTHER EUROPE	AFRICA	OCEANIA
25.7%	12.6	2.1	13.2	14.0	3.9	4.6	2.1	3.2	2.0	11.4	2.8	2.3

Fertilizer Use by Type

POTASH 16.1
PHOSPHATE 23.9
NITROGEN 60.0%

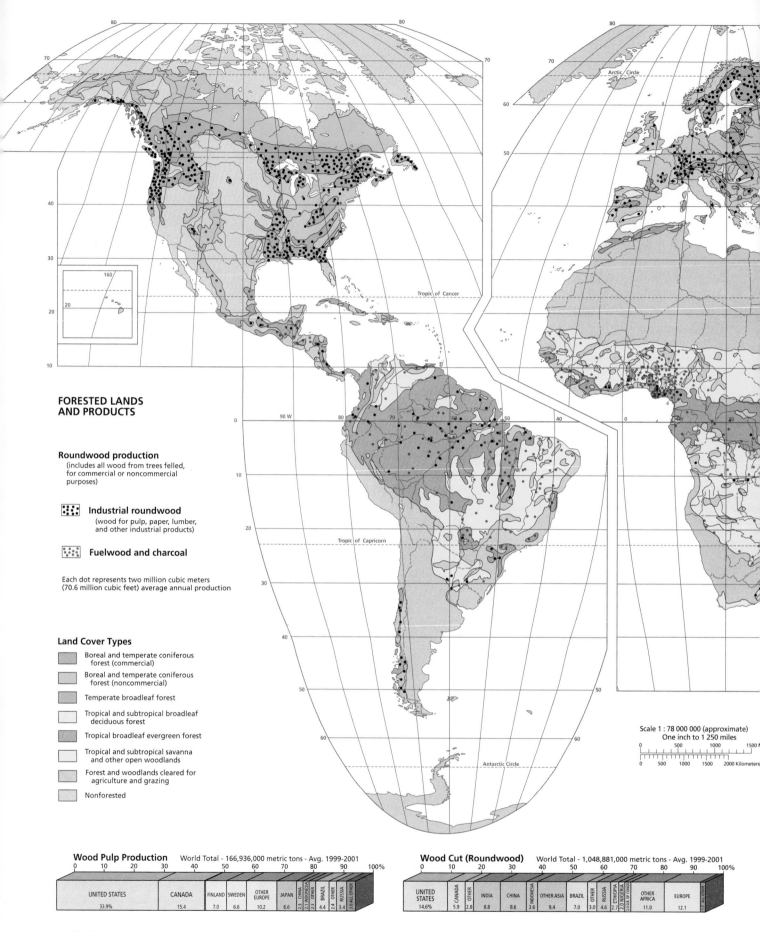

FORESTED LANDS AND PRODUCTS

Roundwood production
(includes all wood from trees felled, for commercial or noncommercial purposes)

Industrial roundwood
(wood for pulp, paper, lumber, and other industrial products)

Fuelwood and charcoal

Each dot represents two million cubic meters (70.6 million cubic feet) average annual production

Land Cover Types

Boreal and temperate coniferous forest (commercial)

Boreal and temperate coniferous forest (noncommercial)

Temperate broadleaf forest

Tropical and subtropical broadleaf deciduous forest

Tropical broadleaf evergreen forest

Tropical and subtropical savanna and other open woodlands

Forest and woodlands cleared for agriculture and grazing

Nonforested

Scale 1 : 78 000 000 (approximate)
One inch to 1 250 miles

0 500 1000 1500 miles
0 500 1000 1500 2000 Kilometers

Wood Pulp Production
World Total - 166,936,000 metric tons - Avg. 1999-2001

UNITED STATES	CANADA	FINLAND	SWEDEN	OTHER EUROPE	JAPAN	CHINA	INDONESIA	OTHER	BRAZIL	RUSSIA	ALL OTHER
33.9%	15.4	7.0	6.6	10.2	6.6	2.3	2.1	2.3	4.4	2.4	3.4	3.0

Wood Cut (Roundwood)
World Total - 1,048,881,000 metric tons - Avg. 1999-2001

UNITED STATES	CANADA	OTHER	INDIA	CHINA	INDONESIA	OTHER ASIA	BRAZIL	OTHER	RUSSIA	ETHIOPIA	NIGERIA	D.R.OF CONGO	OTHER AFRICA	EUROPE	ALL OTHER
14.6%	5.9	2.8	8.8	8.6	3.6	9.4	7.0	3.0	4.6	2.7	2.0	2.0	11.0	12.1	

Tropic of Cancer

Equator

Tropic of Capricorn

Longitude East of Greenwich

Goode's Homolosine Equal Area Projection (Condensed)

Forested Land World Total - 14,940,000 square miles - 2000

RUSSIA 22.0%	BRAZIL 14.1	OTHER S. AMER. 8.8	CANADA 6.3	UNITED STATES 5.8	OTHER 2.2	CHINA 4.2	INDON. 2.7	OTHER ASIA 7.2	AUSTL. 4.0	D.R. OF CONGO 3.5	OTHER AFRICA 13.4	EUROPE 4.9

Reforested Land World Total - 168,000 square miles - 1990-2000

CHINA 41.5%	MALAYSIA 15.2	KAZAKH. 5.5	OTHER ASIA 4.5	UNITED STATES 8.9	BELARUS 5.9	SPAIN 2.0	OTHER EUROPE 9.4	RUSSIA 3.1	ALL OTHER 3.3

Rainforest World Total - 3,877,000 square miles - 2000

BRAZIL 41.2%	PERU 5.6	COLOMBIA 4.1	VENEZUELA 2.5	OTHER S. AMER. 5.3	D.R. OF CONGO 11.0	GABON 2.2	OTHER CONGO 2.1	OTHER AFRICA 6.2	INDONESIA 9.2	OTHER ASIA 5.6	PAPUA NEW GUINEA 2.4 / ALL OTHER 1.8

Deforested Land World Total - 496,000 square miles - 1990-2000

BRAZIL 18.0%	ARG. 2.2	PERU 2.1	OTHER S. AMER. 7.0	INDONESIA 10.2	MYANMAR 4.0	OTHER ASIA 4.2	SUDAN 7.5	ZAMBIA 6.6	D.R. OF CONGO 4.1	NIGERIA 3.1	ZIMBABWE 2.5	COTE D'IVOIRE 2.1	OTHER AFRICA 15.5	MEXICO 4.9	OTHER 2.8 / AUSTRALIA 2.2

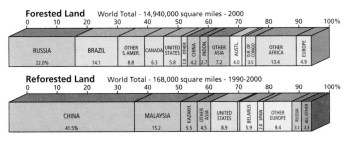
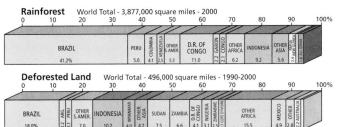

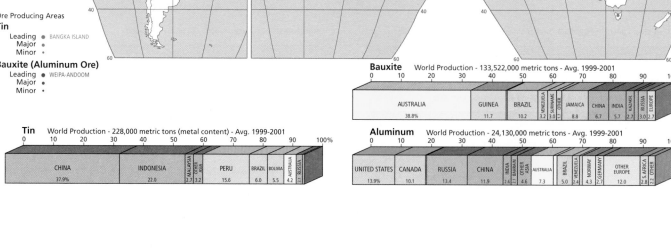

COPPER

NORILSK
ZHEZKAZGAN
SUDBURY-TIMMINS
MORENCI
SOUTHERN PERU
CHUQUICAMATA
ESCONDIDA
EL TENIENTE
MT. ISA

Ore Producing Areas
Leading ● MORENCI
Major ●
Minor •

Copper Reserves
World Total - 940,000,000 metric tons - 2004

| | 0 | 10 | 20 | 30 | 40 | 50 | 60 | 70 | 80 | 90 | 100% |

| CHILE 38.3% | PERU 6.4 | BRAZIL 2.1 | UNITED STATES 7.4 | MEXICO 4.3 | CANADA 2.1 | CHINA 6.7 | INDONESIA 4.0 | KAZAKH. 2.1 | POLAND 5.1 | AUSTRALIA 4.6 | CONGO 4.3 | ZAMBIA 3.7 | RUSSIA 3.2 | ALL OTHER 2.6 |

Copper
World Mine Production - 13,209,000 metric tons (metal content) - Avg. 1999-2001

| 0 | 10 | 20 | 30 | 40 | 50 | 60 | 70 | 80 | 90 | 100% |

| CHILE 34.7% | PERU 4.6 | OTHER 1.6 | UNITED STATES 11.1 | CANADA 4.8 | MEXICO 2.8 | INDONESIA 7.1 | CHINA 4.4 | KAZAKH. 3.2 | OTHER ASIA 4.0 | AUSTRALIA 6.1 | OTHER 1.3 | RUSSIA 4.3 | POLAND 3.5 | OTHER 2.5 | ZAMBIA 2.0 | OTHER 2 |

Refined Copper
World Total - 15,100,000 metric tons - Avg. 1999-2001

| 0 | 10 | 20 | 30 | 40 | 50 | 60 | 70 | 80 | 90 | 100% |

| CHILE 18.2% | PERU 3.0 | UNITED STATES 12.6 | CANADA 3.7 | MEXICO 2.8 | JAPAN 9.3 | CHINA 8.8 | KOREA 3.1 | KAZAKH 2.6 | OTHER ASIA 6.2 | RUSSIA 5.5 | GERMANY 4.7 | POLAND 3.2 | BELGIUM 2.7 | OTHER EUROPE 3.8 | AUSTRALIA 3.2 | ZAMBIA 2.2 |

Copyright by Rand McNally & Co.
Made in U.S.A.
N-GDS10000-F7- -2-3

TIN, BAUXITE

JAMAICA
SANGAREDI
LOS PIJIGUADOS
PORTO TROMBEDAS
SAN RAFAEL
GUANGXI
GEJIU
BANGKA ISLAND
GOVE
WEIPA-ANDOOM
DARLING RANGE

Ore Producing Areas
Tin
Leading ● BANGKA ISLAND
Major ●
Minor •

Bauxite (Aluminum Ore)
Leading ● WEIPA-ANDOOM
Major ●
Minor •

Bauxite
World Production - 133,522,000 metric tons - Avg. 1999-2001

| 0 | 10 | 20 | 30 | 40 | 50 | 60 | 70 | 80 | 90 | 100% |

| AUSTRALIA 38.8% | GUINEA 11.7 | BRAZIL 10.2 | VENEZUELA 3.2 | SURINAME 3.0 | OTHER 1.7 | JAMAICA 8.8 | CHINA 6.7 | INDIA 5.7 | KAZAKH 2.7 | RUSSIA 3.0 | EUROPE 2.7 |

Tin
World Production - 228,000 metric tons (metal content) - Avg. 1999-2001

| 0 | 10 | 20 | 30 | 40 | 50 | 60 | 70 | 80 | 90 | 100% |

| CHINA 37.9% | INDONESIA 22.0 | MALAYSIA 2.7 | OTHER ASIA 3.2 | PERU 15.6 | BRAZIL 6.0 | BOLIVIA 5.5 | AUSTRALIA 2.1 | RUSSIA |

Aluminum
World Production - 24,130,000 metric tons - Avg. 1999-2001

| 0 | 10 | 20 | 30 | 40 | 50 | 60 | 70 | 80 | 90 | 100% |

| UNITED STATES 13.9% | CANADA 10.1 | RUSSIA 13.4 | CHINA 11.9 | INDIA 2.6 | BAHRAIN 2.7 | OTHER ASIA 4.6 | AUSTRALIA 7.3 | BRAZIL 5.0 | VENEZUELA 2.4 | NORWAY 4.3 | GERMANY 2.7 | OTHER EUROPE 12.0 | S. AFRICA 2.8 | OTHER 2.2 |

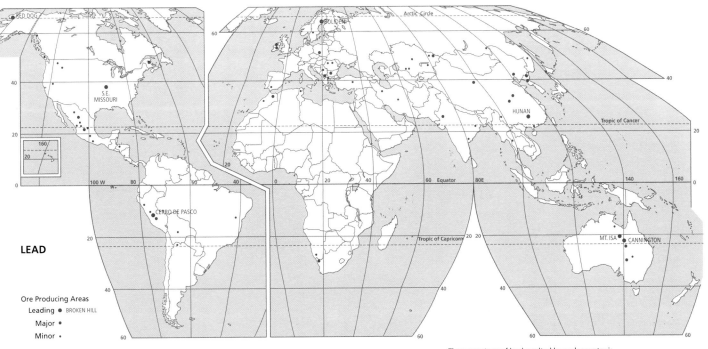

LEAD

Ore Producing Areas

Leading ● BROKEN HILL

Major ●

Minor ·

The percentage of lead smelted by each country is not necessarily identical to its percentage of lead ore production. Some countries, such as Australia, export large amounts of ore to other countries for smelting.

Lead World Mine Production - 3,124,000 metric tons (metal content) - Avg. 1999-2001

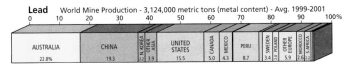

| AUSTRALIA 22.8% | CHINA 19.3 | N. KOREA 2.2 | OTHER ASIA 3.9 | UNITED STATES 15.5 | CANADA 5.0 | MEXICO 4.3 | PERU 8.7 | SWEDEN 3.4 | POLAND 2.0 | OTHER EUROPE 5.9 | MOROCCO 2.6 | S. AFRICA 2.2 |

Lead Smelted* World Production - 6,417,000 metric tons - Avg. 1999-2001

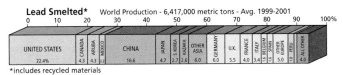

| UNITED STATES 22.4% | CANADA 4.3 | ARUBA 4.3 | MEXICO 2.6 | CHINA 16.6 | JAPAN 4.7 | S. KOREA 2.7 | KAZAKH. 2.6 | OTHER ASIA 6.0 | GERMANY 6.0 | U.K. 5.5 | FRANCE 4.0 | ITALY 3.4 | BELGIUM 1.6 | SPAIN 1.8 | OTHER EUROPE 5.0 | PERU 1.8 | ALL OTHER 4.0 |

*includes recycled materials

Copyright by Rand McNally & Co.
Made in U.S.A.
N-GOS10000-F8- -2- -3

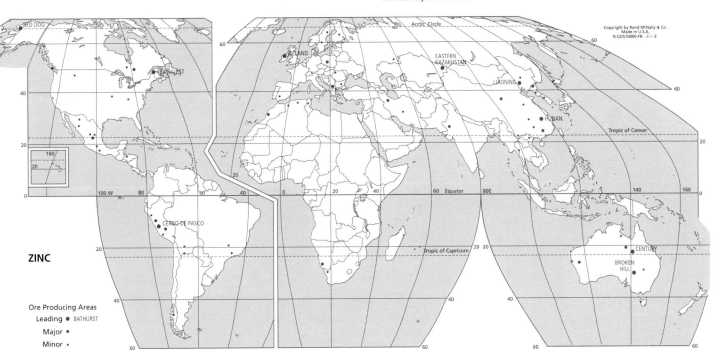

ZINC

Ore Producing Areas

Leading ● BATHURST

Major ●

Minor ·

The percentage of zinc smelted by each country is not necessarily identical to its percentage of zinc ore production. Some countries, such as Australia, export large amounts of ore to other countries for smelting.

Zinc World Mine Production - 8,559,000 metric tons (metal content) - Avg. 1999-2001

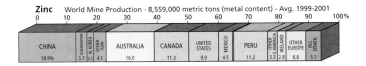

| CHINA 18.9% | KAZAKHSTAN 3.7 | N. KOREA 2.2 | OTHER ASIA 4.1 | AUSTRALIA 16.0 | CANADA 11.3 | UNITED STATES 9.9 | MEXICO 4.5 | PERU 11.2 | OTHER S. AMERICA 3.3 | IRELAND 2.8 | OTHER EUROPE 6.6 | ALL OTHER 5.1 |

Zinc Smelted* World Production - 9,011,000, metric tons - Avg. 1999-2001

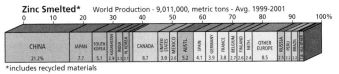

| CHINA 21.2% | JAPAN 7.7 | SOUTH KOREA 5.1 | KAZAKHSTAN 2.9 | INDIA 2.3 | N. KOREA 2.1 | CANADA 8.7 | UNITED STATES 3.9 | MEXICO 2.6 | AUSTL. 5.2 | SPAIN 4.1 | GERMANY 3.9 | FRANCE 3.8 | BELGIUM 2.7 | FINLAND 2.6 | NETH. 2.4 | OTHER EUROPE 8.5 | RUSSIA 2.5 | PERU 2.2 | BRAZIL 2.2 | ALL OTHER 1.6 |

*includes recycled materials

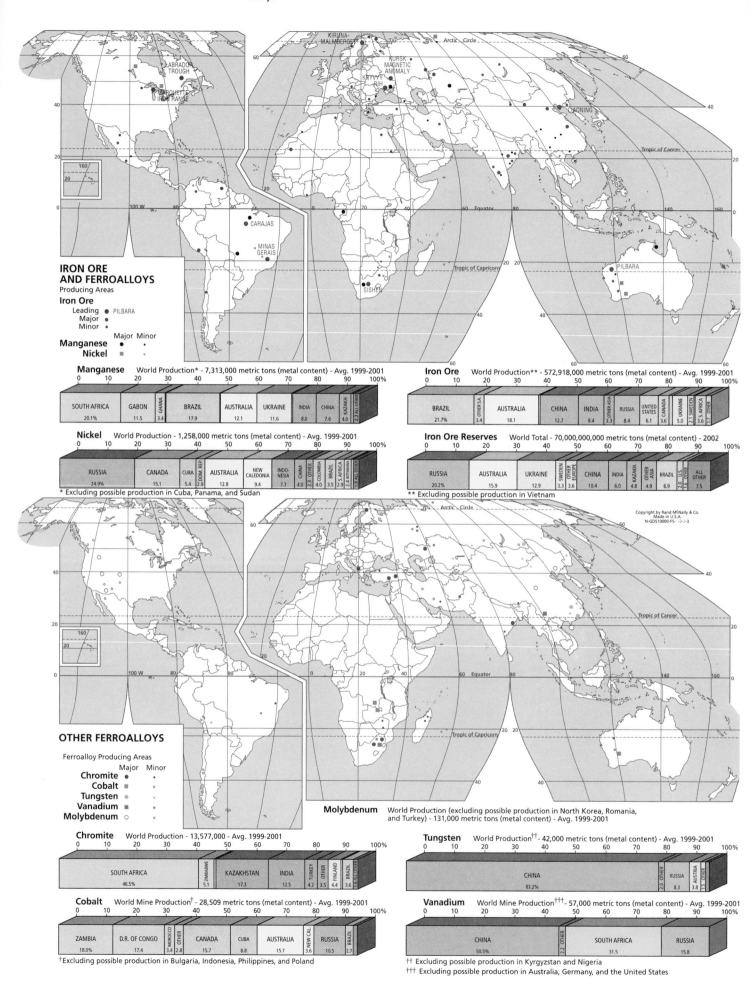

IRON ORE AND FERROALLOYS

Producing Areas

Iron Ore
Leading ● PILBARA
Major ●
Minor ·

Major Minor
Manganese ● ·
Nickel ■ ·

OTHER FERROALLOYS

Ferroalloy Producing Areas

Major Minor
Chromite ● ·
Cobalt ■ ·
Tungsten ● ·
Vanadium ■ ·
Molybdenum ○ ○

Manganese World Production* - 7,313,000 metric tons (metal content) - Avg. 1999-2001

| SOUTH AFRICA 20.1% | GABON 11.5 | GHANA 3.4 | BRAZIL 17.9 | AUSTRALIA 12.1 | UKRAINE 11.6 | INDIA 8.0 | CHINA 7.6 | KAZAKH. 4.0 | ALL OTHER 2.2 |

Nickel World Production - 1,258,000 metric tons (metal content) - Avg. 1999-2001

| RUSSIA 24.9% | CANADA 15.1 | CUBA 5.4 | DOM. REP. | AUSTRALIA 12.8 | NEW CALEDONIA 9.4 | INDO-NESIA 7.7 | CHINA 4.0 | OTHER 2.3 | COLOMBIA 4.0 | BRAZIL 3.5 | S. AFRICA 2.4 | BOTSWANA 2.9 | ALL OTHER 1.9 |

* Excluding possible production in Cuba, Panama, and Sudan

Iron Ore World Production** - 572,918,000 metric tons (metal content) - Avg. 1999-2001

| BRAZIL 21.7% | OTHER S.A. 3.4 | AUSTRALIA 18.1 | CHINA 12.7 | INDIA 8.4 | OTHER ASIA 3.3 | RUSSIA 8.4 | UNITED STATES 6.1 | CANADA 3.6 | UKRAINE 2.1 | SWEDEN 1.7 | S. AFRICA 3.6 | OTHER |

Iron Ore Reserves World Total - 70,000,000,000 metric tons (metal content) - 2002

| RUSSIA 20.2% | AUSTRALIA 15.9 | UKRAINE 12.9 | SWEDEN 3.3 | OTHER EUROPE 3.6 | CHINA 10.4 | INDIA 6.0 | KAZAKH. 4.8 | OTHER ASIA 4.9 | BRAZIL 6.9 | U.S. 2.0 | OTHER 1.6 | ALL OTHER 7.5 |

** Excluding possible production in Vietnam

Copyright by Rand McNally & Co.
Made in U.S.A.
N-GDS10000-F5- -2-2-3

Chromite World Production - 13,577,000 - Avg. 1999-2001

| SOUTH AFRICA 46.5% | ZIMBABWE 5.1 | KAZAKHSTAN 17.3 | INDIA 12.5 | TURKEY 4.2 | OTHER 3.5 | FINLAND 4.4 | BRAZIL 2.4 | ALL OTHER 1.6 |

Cobalt World Mine Production† - 28,509 metric tons (metal content) - Avg. 1999-2001

| ZAMBIA 18.0% | D.R. OF CONGO 17.4 | MOROCCO 3.4 | OTHER 2.8 | CANADA 15.7 | CUBA 8.8 | AUSTRALIA 15.7 | NEW CAL 3.6 | RUSSIA 10.5 | BRAZIL 2.7 |

†Excluding possible production in Bulgaria, Indonesia, Philippines, and Poland

Molybdenum World Production (excluding possible production in North Korea, Romania, and Turkey) - 131,000 metric tons (metal content) - Avg. 1999-2001

Tungsten World Production†† - 42,000 metric tons (metal content) - Avg. 1999-2001

| CHINA 83.2% | OTHER 2.0 | RUSSIA 8.3 | AUSTRIA 3.8 | OTHER 1.5 |

Vanadium World Mine Production††† - 57,000 metric tons (metal content) - Avg. 1999-2001

| CHINA 50.5% | OTHER 2.2 | SOUTH AFRICA 31.5 | RUSSIA 15.8 |

†† Excluding possible production in Kyrgyzstan and Nigeria
††† Excluding possible production in Australia, Germany, and the United States

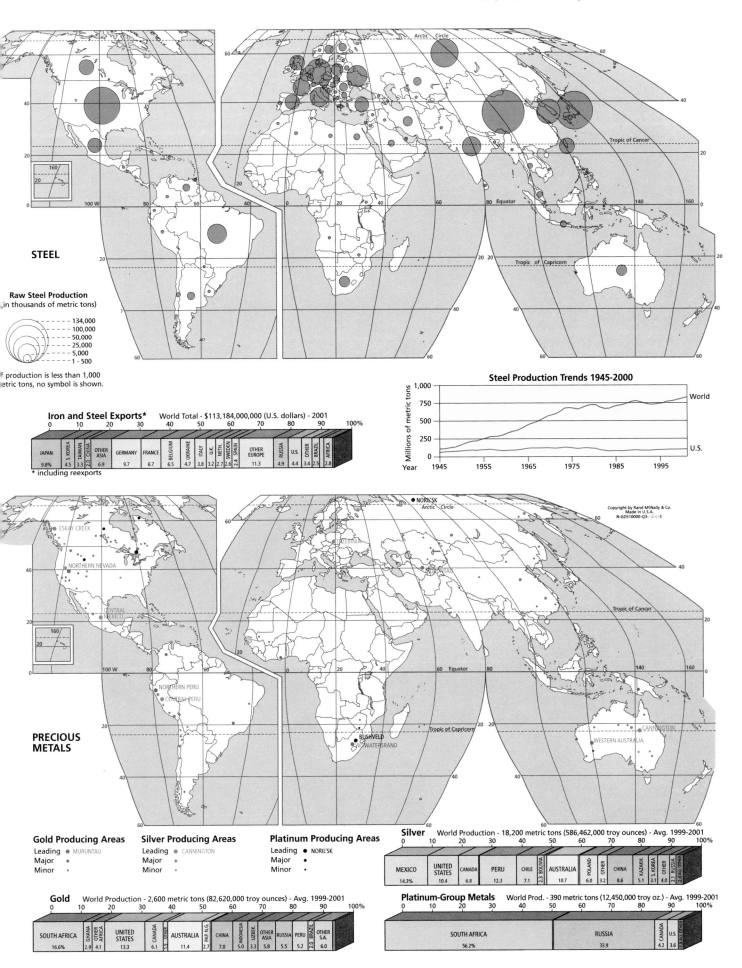

STEEL

Raw Steel Production
(in thousands of metric tons)

- 134,000
- 100,000
- 50,000
- 25,000
- 5,000
- 1 - 500

If production is less than 1,000 metric tons, no symbol is shown.

Iron and Steel Exports* World Total - $113,184,000,000 (U.S. dollars) - 2001

	0	10	20	30	40	50	60	70	80	90	100%

JAPAN	S. KOREA	TAIWAN	CHINA	OTHER ASIA	GERMANY	FRANCE	BELGIUM	ITALY	U.K.	NETH.	SWEDEN	SPAIN	OTHER EUROPE	RUSSIA	U.S.	OTHER	BRAZIL	AFRICA
9.8%	4.5	3.3	2.0	6.9	9.7	6.7	6.5	4.7	3.8	3.2	2.7	2.6	2.4	11.3	4.9	4.4	3.4	2.5	2.8

* including reexports

Steel Production Trends 1945-2000

Millions of metric tons — 1,000 / 750 / 500 / 250

World

U.S.

Year 1945   1955   1965   1975   1985   1995

PRECIOUS METALS

NORIL'SK
Arctic Circle

ESKAY CREEK

NORTHERN NEVADA

CENTRAL MEXICO

TAZEBINIA

MURUNTAU

NORTHERN PERU

CENTRAL PERU

BUSHVELD
WITWATERSRAND

CANNINGTON

WESTERN AUSTRALIA

Gold Producing Areas

Leading ● MURUNTAU
Major ●
Minor ·

Silver Producing Areas

Leading ● CANNINGTON
Major ●
Minor ·

Platinum Producing Areas

Leading ● NORIL'SK
Major ●
Minor ·

Silver World Production - 18,200 metric tons (586,462,000 troy ounces) - Avg. 1999-2001

	0	10	20	30	40	50	60	70	80	90	100%

MEXICO	UNITED STATES	CANADA	PERU	CHILE	BOLIVIA	AUSTRALIA	POLAND	OTHER	CHINA	KAZAKH.	S. KOREA	OTHER	RUSSIA	ALL OTHER
14.3%	10.4	6.8	12.3	7.1	2.3	10.7	6.0	3.2	8.6	5.1	3.1	4.0	2.1	2.1

Gold World Production - 2,600 metric tons (82,620,000 troy ounces) - Avg. 1999-2001

	0	10	20	30	40	50	60	70	80	90	100%

SOUTH AFRICA	GHANA	OTHER AFRICA	UNITED STATES	CANADA	OTHER	AUSTRALIA	PAP. N.G.	CHINA	INDONESIA	UZBEK.	OTHER ASIA	RUSSIA	PERU	BRAZIL	OTHER S.A.
16.6%	2.9	4.1	13.3	6.1	1.5	11.4	2.7	7.0	5.0	3.3	5.8	5.5	5.2	2.0	6.0

Platinum-Group Metals World Prod. - 390 metric tons (12,450,000 troy oz.) - Avg. 1999-2001

	0	10	20	30	40	50	60	70	80	90	100%

SOUTH AFRICA	RUSSIA	CANADA	U.S.	ALL OTHER
56.2%	33.9	4.2	3.6	1.8

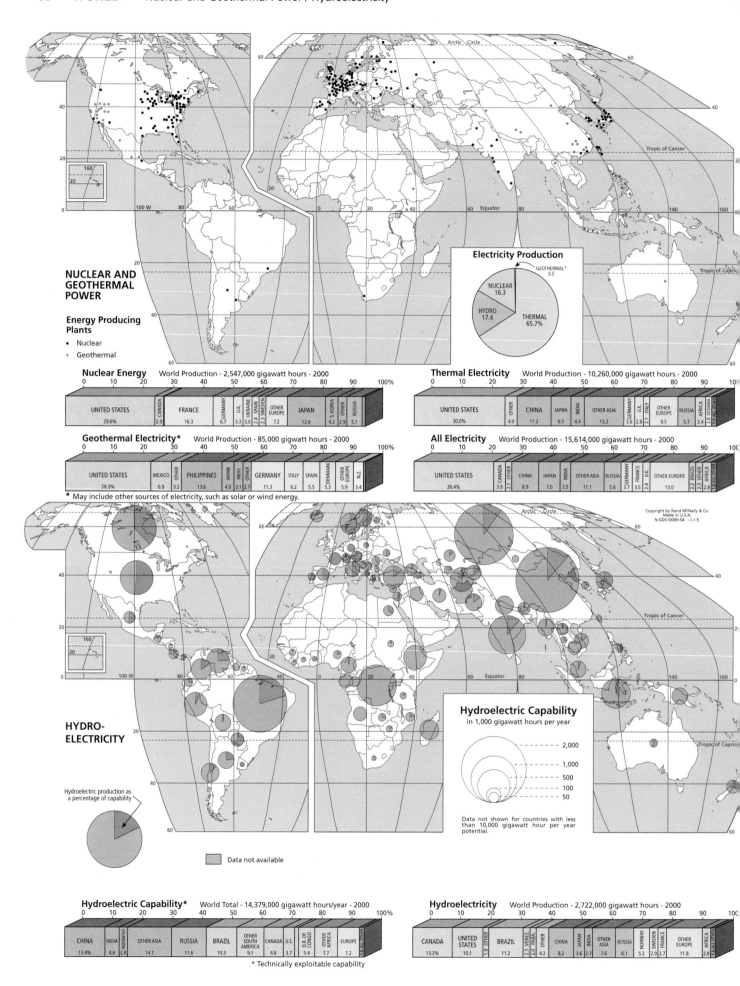

NUCLEAR AND GEOTHERMAL POWER

Energy Producing Plants

- Nuclear
- Geothermal

Electricity Production

- GEOTHERMAL* 0.5
- NUCLEAR 16.3
- HYDRO 17.4
- THERMAL 65.7%

Nuclear Energy World Production - 2,547,000 gigawatt hours - 2000

| UNITED STATES 29.6% | CANADA 2.9 | FRANCE 16.3 | GERMANY 6.7 | U.K. 3.3 | UKRAINE 2.4 | SPAIN 2.3 | SWEDEN | OTHER EUROPE 7.2 | JAPAN 12.6 | S. KOREA 4.3 | OTHER 2.9 | RUSSIA 5.1 |

Geothermal Electricity* World Production - 85,000 gigawatt hours - 2000

| UNITED STATES 28.3% | MEXICO 6.9 | OTHER 3.2 | PHILIPPINES 13.6 | JAPAN 4.0 | INDO. 3.1 | OTHER 2.7 | GERMANY 11.3 | ITALY 6.2 | SPAIN 5.5 | DENMARK 5.3 | OTHER EUROPE 5.9 | N.Z. 3.4 |

* May include other sources of electricity, such as solar or wind energy.

Thermal Electricity World Production - 10,260,000 gigawatt hours - 2000

| UNITED STATES 30.0% | OTHER 4.0 | CHINA 11.2 | JAPAN 6.5 | INDIA 4.4 | OTHER ASIA 13.2 | GERMANY 3.6 | U.K. 2.8 | ITALY 2.1 | OTHER EUROPE 9.5 | RUSSIA 5.7 | AFRICA 3.4 | OCEANIA 2.0 | ALL OTHER |

All Electricity World Production - 15,614,000 gigawatt hours - 2000

| UNITED STATES 26.4% | CANADA 3.8 | OTHER 2.1 | CHINA 8.9 | JAPAN 7.0 | INDIA 3.5 | OTHER ASIA 11.1 | RUSSIA 5.6 | GERMANY 3.7 | FRANCE 3.5 | U.K. 2.4 | OTHER EUROPE 13.0 | BRAZIL 2.2 | AFRICA 2.3 | ALL OTHER 2.8 |

HYDRO-ELECTRICITY

Hydroelectric production as a percentage of capability

Hydroelectric Capability
in 1,000 gigawatt hours per year

- 2,000
- 1,000
- 500
- 100
- 50

Data not shown for countries with less than 10,000 gigawatt hour per year potential.

Copyright by Rand McNally & Co.
Made in U.S.A.
N-GDS10000-S4- -3-4-5

Data not available

Hydroelectric Capability* World Total - 14,379,000 gigawatt hours/year - 2000

| CHINA 13.4% | INDIA 4.6 | INDONESIA 2.8 | OTHER ASIA 14.7 | RUSSIA 11.6 | BRAZIL 10.3 | OTHER SOUTH AMERICA 9.1 | CANADA 6.6 | U.S. 3.7 | D.R. OF CONGO 5.4 | OTHER AFRICA 7.7 | EUROPE 7.2 | ALL OTHER |

* Technically exploitable capability

Hydroelectricity World Production - 2,722,000 gigawatt hours - 2000

| CANADA 13.2% | UNITED STATES 10.1 | OTHER 1.9 | BRAZIL 11.2 | VENEZ. PARA. 2.3 | OTHER 2.0 | OTHER 4.2 | CHINA 8.2 | JAPAN 3.6 | INDIA 2.7 | OTHER ASIA 7.6 | RUSSIA 6.1 | NORWAY 5.2 | SWEDEN 2.9 | FRANCE 2.7 | OTHER EUROPE 11.8 | AFRICA 2.3 | ALL OTHER |

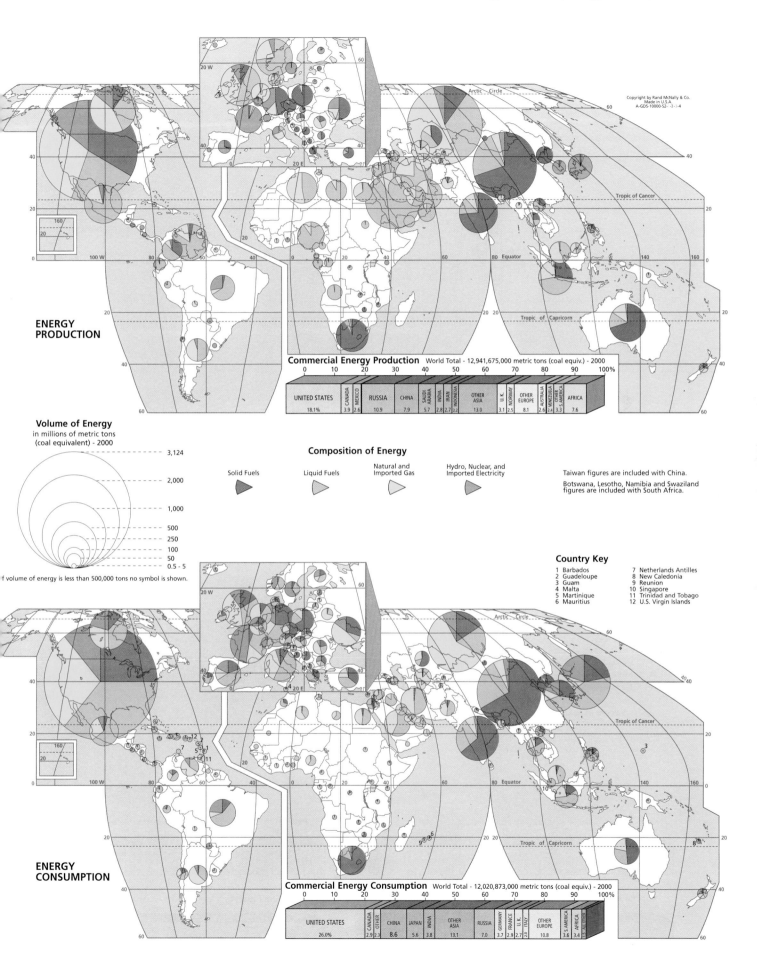

ENERGY
PRODUCTION

Commercial Energy Production World Total - 12,941,675,000 metric tons (coal equiv.) - 2000

0	10		20	30	40	50		60	70		80		90		100%	

UNITED STATES	CANADA	MEXICO	RUSSIA	CHINA	SAUDI ARABIA	INDIA	IRAN	INDONESIA	OTHER ASIA	U.K.	NORWAY	OTHER EUROPE	AUSTRALIA	VENEZUELA	OTHER S. AMERICA	AFRICA
18.1%	3.9	2.6	10.9	7.9	5.7	2.8	2.7	2.2	13.0	3.1	2.5	8.1	2.6	2.4	3.3	7.6

Volume of Energy
in millions of metric tons
(coal equivalent) - 2000

- 3,124
- 2,000
- 1,000
- 500
- 250
- 100
- 50
- 0.5 - 5

If volume of energy is less than 500,000 tons no symbol is shown.

Composition of Energy

Solid Fuels   Liquid Fuels   Natural and Imported Gas   Hydro, Nuclear, and Imported Electricity

Taiwan figures are included with China.

Botswana, Lesotho, Namibia and Swaziland figures are included with South Africa.

Country Key

1 Barbados
2 Guadeloupe
3 Guam
4 Malta
5 Martinique
6 Mauritius

7 Netherlands Antilles
8 New Caledonia
9 Reunion
10 Singapore
11 Trinidad and Tobago
12 U.S. Virgin Islands

ENERGY
CONSUMPTION

Commercial Energy Consumption World Total - 12,020,873,000 metric tons (coal equiv.) - 2000

0	10	20	30	40	50	60		70	80			90		100%

UNITED STATES	CANADA	OTHER	CHINA	JAPAN	INDIA	OTHER ASIA	RUSSIA	GERMANY	FRANCE	U.K.	ITALY	OTHER EUROPE	S. AMERICA	AFRICA	AUSTRALIA
26.0%	2.9	2.3	8.6	5.6	3.8	13.1	7.0	3.7	2.9	2.7	2.0	10.8	3.6	3.4	

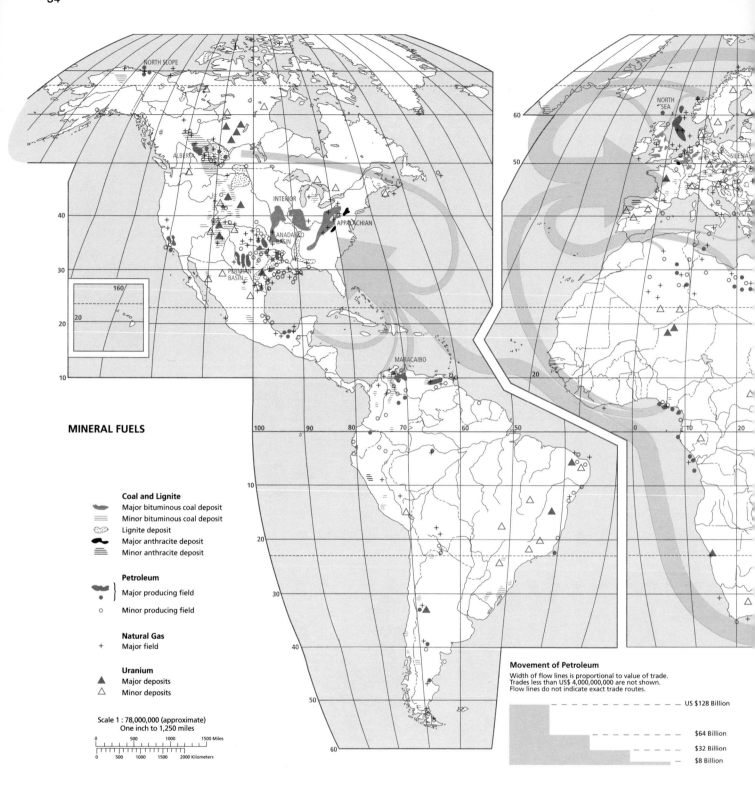

MINERAL FUELS

Coal and Lignite
- Major bituminous coal deposit
- Minor bituminous coal deposit
- Lignite deposit
- Major anthracite deposit
- Minor anthracite deposit

Petroleum
- Major producing field
- Minor producing field

Natural Gas
- + Major field

Uranium
- ▲ Major deposits
- △ Minor deposits

Scale 1 : 78,000,000 (approximate)
One inch to 1,250 miles

0   500   1000   1500 Miles

0   500   1000   1500   2000 Kilometers

Movement of Petroleum
Width of flow lines is proportional to value of trade.
Trades less than US$ 4,000,000,000 are not shown.
Flow lines do not indicate exact trade routes.

- US $128 Billion
- $64 Billion
- $32 Billion
- $8 Billion

Map labels: NORTH SLOPE, ALBERTA, INTERIOR, ANADARKO BASIN, APPALACHIAN, PERMIAN BASIN, MARACAIBO, NORTH SEA, SILESIA

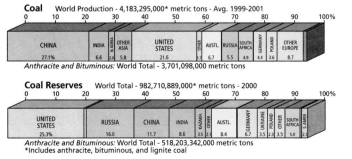

Coal World Production - 4,183,295,000* metric tons - Avg. 1999-2001

0	10	20	30	40	50	60	70	80	90	100%

CHINA	INDIA	N. KOREA	OTHER ASIA	UNITED STATES	OTHER	AUSTL.	RUSSIA	SOUTH AFRICA	GERMANY	POLAND	OTHER EUROPE
27.1%	6.6	2.0	5.8	21.6	1.7	6.7	5.5	4.9	4.4	3.6	8.7

Anthracite and Bituminous: World Total - 3,701,098,000 metric tons

Coal Reserves World Total - 982,710,889,000* metric tons - 2000

0	10	20	30	40	50	60	70	80	90	100%

UNITED STATES	RUSSIA	CHINA	INDIA	KAZAKH.	OTHER	AUSTL.	GERMANY	UKRAINE	POLAND	OTHER	SOUTH AFRICA	S. AMER.
25.3%	16.0	11.7	8.6	3.5	2.1	8.4	6.7	3.5	2.3	3.5	5.0	2.1

Anthracite and Bituminous: World Total - 518,203,342,000 metric tons
*Includes anthracite, bituminous, and lignite coal

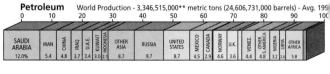

Petroleum World Production - 3,346,515,000** metric tons (24,606,731,000 barrels) - Avg. 199

0	10	20	30	40	50	60	70	80	90	100

SAUDI ARABIA	IRAN	CHINA	IRAQ	U.A.E.	KUWAIT	INDONESIA	OTHER ASIA	RUSSIA	UNITED STATES	MEXICO	CANADA	NORWAY	U.K.	VENEZ.	OTHER S. AMERICA	NIGERIA	LIBYA	OTHER AFRICA
12.0%	5.4	4.8	3.7	3.4	3.0	2.1	8.7	9.7	8.7	4.5	2.9	4.6	3.6	4.4	4.8	3.2	2.6	5.8

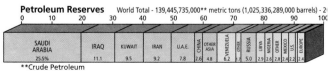

Petroleum Reserves World Total - 139,445,735,000** metric tons (1,025,336,289,000 barrels) - 2

0	10	20	30	40	50	60	70	80	90	100

SAUDI ARABIA	IRAQ	KUWAIT	IRAN	U.A.E.	CHINA	OTHER ASIA	VENEZUELA	OTHER	RUSSIA	LIBYA	NIGERIA	MEXICO	U.S.	EUROPE
25.5%	11.1	9.5	9.2	7.8	2.6	4.8	6.2	0.7	5.0	2.6	2.8	2.2	2.4	2.4

**Crude Petroleum

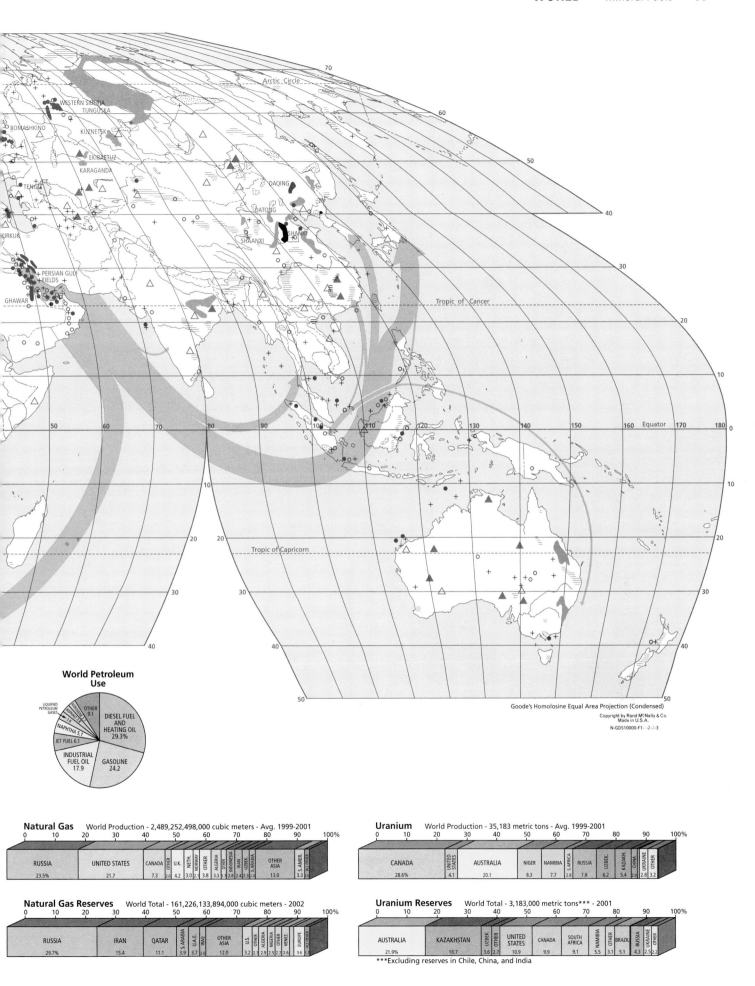

Arctic Circle

WESTERN SIBERIA
TUNGUSKA
ROMASHKINO
KUZNETSK
EKIBASTUZ
KARAGANDA
TENGIZ
DAQING
DATONG
SHANXI
SHAANXI
KIRKUK
PERSIAN GULF
FIELDS
GHAWAR

Tropic of Cancer

Equator

Tropic of Capricorn

Goode's Homolosine Equal Area Projection (Condensed)
Copyright by Rand McNally & Co.
Made in U.S.A.
N-GDS10000-F1- -2-2-3

World Petroleum Use

- DIESEL FUEL AND HEATING OIL 29.3%
- GASOLINE 24.2
- INDUSTRIAL FUEL OIL 17.9
- JET FUEL 6.1
- NAPHTHA 5.7
- LIQUIFIED PETROLEUM GASES
- KEROSENE
- ASPHALT
- OTHER 9.1

Natural Gas — World Production - 2,489,252,498,000 cubic meters - Avg. 1999-2001

0	10	20	30	40	50	60	70	80	90	100%

| RUSSIA 23.5% | UNITED STATES 21.7 | CANADA 7.3 | OTHER 2.0 | U.K. 4.2 | NETH. 3.0 | NORWAY 2.1 | OTHER 3.8 | ALGERIA 3.3 | OTHER 1.7 | INDONESIA 2.8 | IRAN 2.4 | UZBEK. 2.1 | S. ARABIA 2.0 | OTHER ASIA 13.0 | S. AMER. 3.3 | ALL OTHER |

Natural Gas Reserves — World Total - 161,226,133,894,000 cubic meters - 2002

0	10	20	30	40	50	60	70	80	90	100%

| RUSSIA 29.7% | IRAN 15.4 | QATAR 11.1 | S. ARABIA 3.9 | U.A.E. 3.7 | IRAQ 2.0 | OTHER ASIA 12.0 | U.S. 3.2 | OTHER 2.1 | ALGERIA 2.9 | NIGERIA 2.5 | OTHER 2.2 | VENEZ. 2.6 | EUROPE 3.6 | ALL OTHER |

Uranium — World Production - 35,183 metric tons - Avg. 1999-2001

0	10	20	30	40	50	60	70	80	90	100%

| CANADA 28.6% | UNITED STATES 4.1 | AUSTRALIA 20.1 | NIGER 8.3 | NAMIBIA 7.7 | S. AFRICA 2.8 | RUSSIA 7.8 | UZBEK. 6.2 | KAZAKH. 5.4 | CHINA 2.9 | UKRAINE 2.8 | OTHER 3.2 |

Uranium Reserves — World Total - 3,183,000 metric tons*** - 2001

0	10	20	30	40	50	60	70	80	90	100%

| AUSTRALIA 21.9% | KAZAKHSTAN 18.7 | UZBEK. 3.6 | OTHER 2.7 | UNITED STATES 10.9 | CANADA 9.9 | SOUTH AFRICA 9.1 | NAMIBIA 5.5 | OTHER 3.1 | BRAZIL 5.1 | RUSSIA 4.3 | UKRAINE 2.5 | OTHER 2.2 |

***Excluding reserves in Chile, China, and India

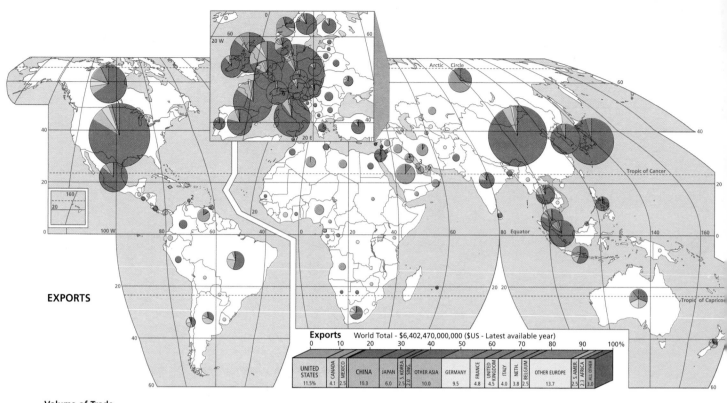

EXPORTS

Exports	World Total - $6,402,470,000,000 ($US - Latest available year)																

0	10		20	30	40	50	60	70		80		90		100%

UNITED STATES	CANADA	MEXICO	CHINA	JAPAN	S. KOREA	SING.	OTHER ASIA	GERMANY	FRANCE	UNITED KINGDOM	ITALY	NETH.	BELGIUM	OTHER EUROPE	S. AMER.	AFRICA	ALL OTHER
11.5%	4.1	2.5	10.3	6.0	2.5	2.0	10.0	9.5	4.8	4.5	4.0	3.8	2.5	13.7	2.5	2.3	3.0

Volume of Trade
in billions of U.S. dollars - latest available year

- 1,200
- 500
- 200
- 100
- 50
- 20
- 10
- 1 - 2

If volume of trade is less than 15 billion dollars, color indicates major class only. If no symbol is shown, volume of trade is less than 1 billion dollars.

Composition of Trade

Manufactured Articles	Food, Beverage & Tobacco	Raw Materials	Fuel & Related Products	All Other or Undifferentiated

Taiwan figures are included with China.
Puerto Rico figures are included with the United States.

Data not available

Country Key
1 Andorra	6 Liechtenstein
2 Aruba	7 Malta
3 Bahrain	8 Martinique
4 Gaza Strip and West Bank	9 Netherlands Antilles
5 Guadeloupe	10 Qatar

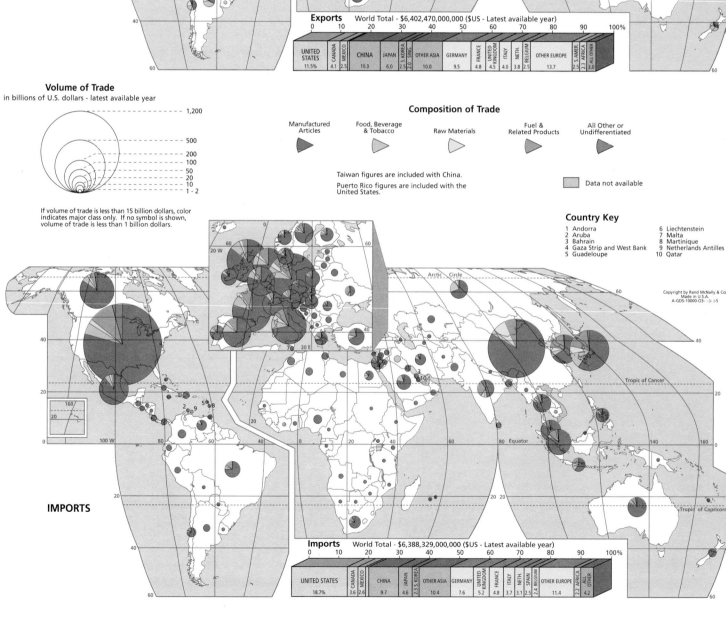

IMPORTS

Imports	World Total - $6,388,329,000,000 ($US - Latest available year)

0	10	20	30	40	50	60	70	80	90	100%

UNITED STATES	CANADA	MEXICO	CHINA	JAPAN	S. KOREA	OTHER ASIA	GERMANY	UNITED KINGDOM	FRANCE	ITALY	NETH.	SPAIN	BELGIUM	OTHER EUROPE	AFRICA	ALL OTHER
18.7%	3.6	2.6	9.7	4.6	2.3	10.4	7.6	5.2	4.8	3.7	3.1	2.5	2.4	11.4	2.2	4.2

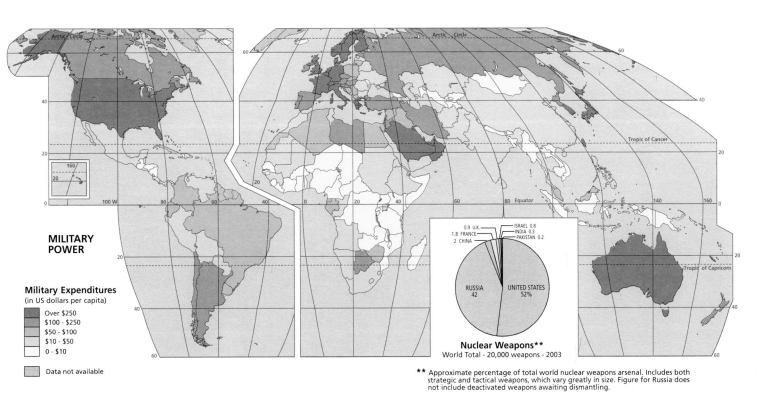

MILITARY POWER

Military Expenditures
(in US dollars per capita)

- Over $250
- $100 - $250
- $50 - $100
- $10 - $50
- 0 - $10

- Data not available

Nuclear Weapons**
World Total - 20,000 weapons - 2003

0.9 U.K.
ISRAEL 0.8
1.8 FRANCE
INDIA 0.3
2 CHINA
PAKISTAN 0.2

RUSSIA 42
UNITED STATES 52%

** Approximate percentage of total world nuclear weapons arsenal. Includes both strategic and tactical weapons, which vary greatly in size. Figure for Russia does not include deactivated weapons awaiting dismantling.

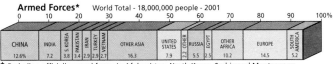

Armed Forces* World Total - 18,000,000 people - 2001

CHINA	INDIA	S. KOREA	PAKISTAN	IRAN	TURKEY	VIETNAM	OTHER ASIA	UNITED STATES	OTHER	RUSSIA	EGYPT	OTHER AFRICA	EUROPE	SOUTH AMERICA
12.6%	7.2	3.8	3.4	2.9	2.9	2.7	16.3	7.9	2.2	5.5	2.5	10.2	14.5	5.2

* Excluding officially armed forces in Afghanistan, North Korea, Serbia and Montenegro, Somalia, and Taiwan.

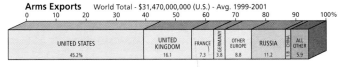

Arms Exports World Total - $31,470,000,000 (U.S.) - Avg. 1999-2001

UNITED STATES	UNITED KINGDOM	FRANCE	GERMANY	OTHER EUROPE	RUSSIA	CHINA	ALL OTHER
45.2%	16.1	7.3	3.8	8.8	11.2	1.8	5.9

Copyright by Rand McNally & Co.
Made in U.S.A.
A-GDS-10000-Y6- -:-:-1

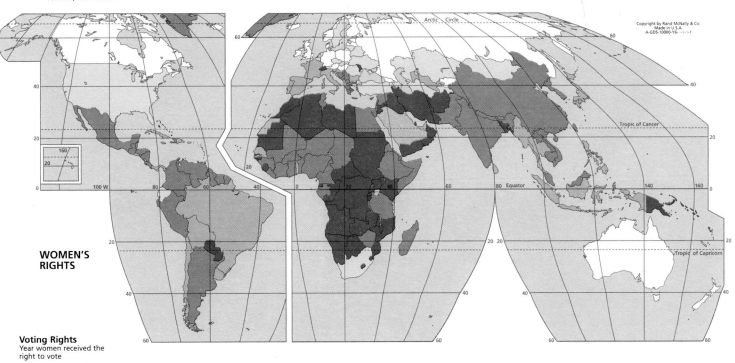

WOMEN'S RIGHTS

Voting Rights
Year women received the right to vote

- After 1960
- 1946 - 1960
- 1931 - 1945
- 1919 - 1930
- Before 1919

- Not Applicable*

*Women are not allowed to vote in Kuwait. Neither women nor men are allowed to vote in Brunei, Saudi Arabia, United Arab Emirates, or Western Sahara.

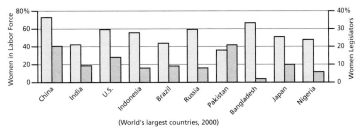

Women's Economic Activity and Legislative Participation Rates

- Percentage of women aged 15 and above in the economically active labor force
- Percentage of seats in national legislature held by women

(World's largest countries, 2000)

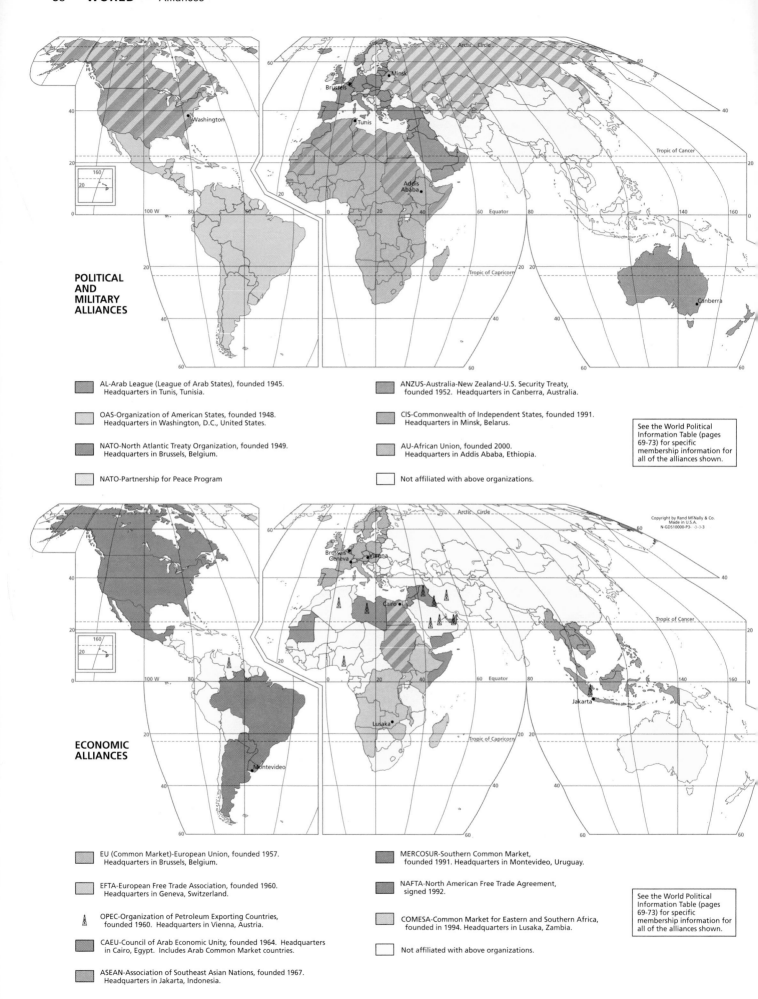

POLITICAL AND MILITARY ALLIANCES

AL-Arab League (League of Arab States), founded 1945. Headquarters in Tunis, Tunisia.

OAS-Organization of American States, founded 1948. Headquarters in Washington, D.C., United States.

NATO-North Atlantic Treaty Organization, founded 1949. Headquarters in Brussels, Belgium.

NATO-Partnership for Peace Program

ANZUS-Australia-New Zealand-U.S. Security Treaty, founded 1952. Headquarters in Canberra, Australia.

CIS-Commonwealth of Independent States, founded 1991. Headquarters in Minsk, Belarus.

AU-African Union, founded 2000. Headquarters in Addis Ababa, Ethiopia.

Not affiliated with above organizations.

See the World Political Information Table (pages 69-73) for specific membership information for all of the alliances shown.

ECONOMIC ALLIANCES

EU (Common Market)-European Union, founded 1957. Headquarters in Brussels, Belgium.

EFTA-European Free Trade Association, founded 1960. Headquarters in Geneva, Switzerland.

OPEC-Organization of Petroleum Exporting Countries, founded 1960. Headquarters in Vienna, Austria.

CAEU-Council of Arab Economic Unity, founded 1964. Headquarters in Cairo, Egypt. Includes Arab Common Market countries.

ASEAN-Association of Southeast Asian Nations, founded 1967. Headquarters in Jakarta, Indonesia.

MERCOSUR-Southern Common Market, founded 1991. Headquarters in Montevideo, Uruguay.

NAFTA-North American Free Trade Agreement, signed 1992.

COMESA-Common Market for Eastern and Southern Africa, founded in 1994. Headquarters in Lusaka, Zambia.

Not affiliated with above organizations.

See the World Political Information Table (pages 69-73) for specific membership information for all of the alliances shown.

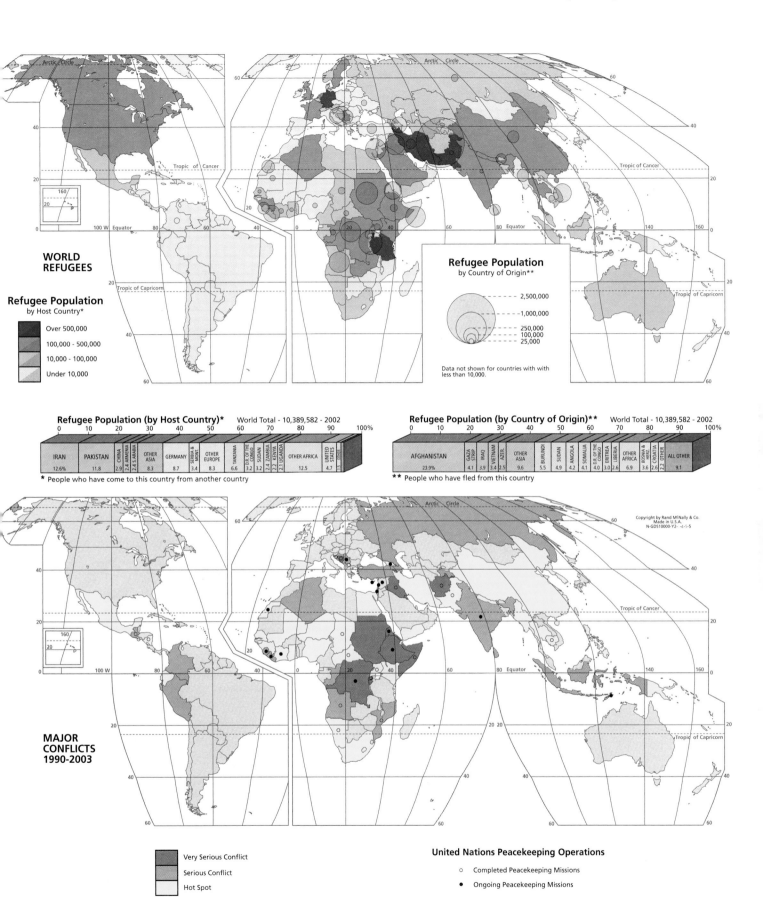

WORLD REFUGEES

Refugee Population
by Host Country*

■	Over 500,000
■	100,000 - 500,000
■	10,000 - 100,000
☐	Under 10,000

Refugee Population
by Country of Origin**

- - - - - 2,500,000
- - - - - 1,000,000

- - - - - 250,000
- - - - - 100,000
- - - - - 25,000

Data not shown for countries with with less than 10,000.

Refugee Population (by Host Country)*   World Total - 10,389,582 - 2002

| 0 | 10 | 20 | 30 | 40 | 50 | 60 | 70 | 80 | 90 | 100% |

IRAN	PAKISTAN	CHINA	ARMENIA	S. ARABIA	OTHER ASIA	GERMANY	SERBIA & MONT.	OTHER EUROPE	TANZANIA	D.R. OF THE CONGO	SUDAN	ZAMBIA	KENYA	UGANDA	OTHER AFRICA	UNITED STATES	OTHER
12.6%	11.8	2.9	2.4	2.4	8.3	8.7	3.4	8.3	6.6	3.2	3.2	2.4	2.2	2.1	12.5	4.7	1.5

* People who have come to this country from another country

Refugee Population (by Country of Origin)**   World Total - 10,389,582 - 2002

| 0 | 10 | 20 | 30 | 40 | 50 | 60 | 70 | 80 | 90 | 100% |

AFGHANISTAN	GAZA STRIP	IRAQ	VIETNAM	AZER.	OTHER ASIA	BURUNDI	SUDAN	ANGOLA	SOMALIA	D.R. OF THE CONGO	ERITREA	LIBERIA	OTHER AFRICA	BOSNIA & HERZ.	CROATIA	OTHER	ALL OTHER
23.9%	4.1	3.9	3.4	2.5	9.6	5.5	4.9	4.2	4.1	4.0	3.0	2.6	6.9	3.6	2.6	2.2	9.1

** People who have fled from this country

Copyright by Rand McNally & Co.
Made in U.S.A.
N-GD510000-Y2- -4-5-5

MAJOR CONFLICTS 1990-2003

■	Very Serious Conflict
■	Serious Conflict
☐	Hot Spot

United Nations Peacekeeping Operations

- ○ Completed Peacekeeping Missions
- ● Ongoing Peacekeeping Missions

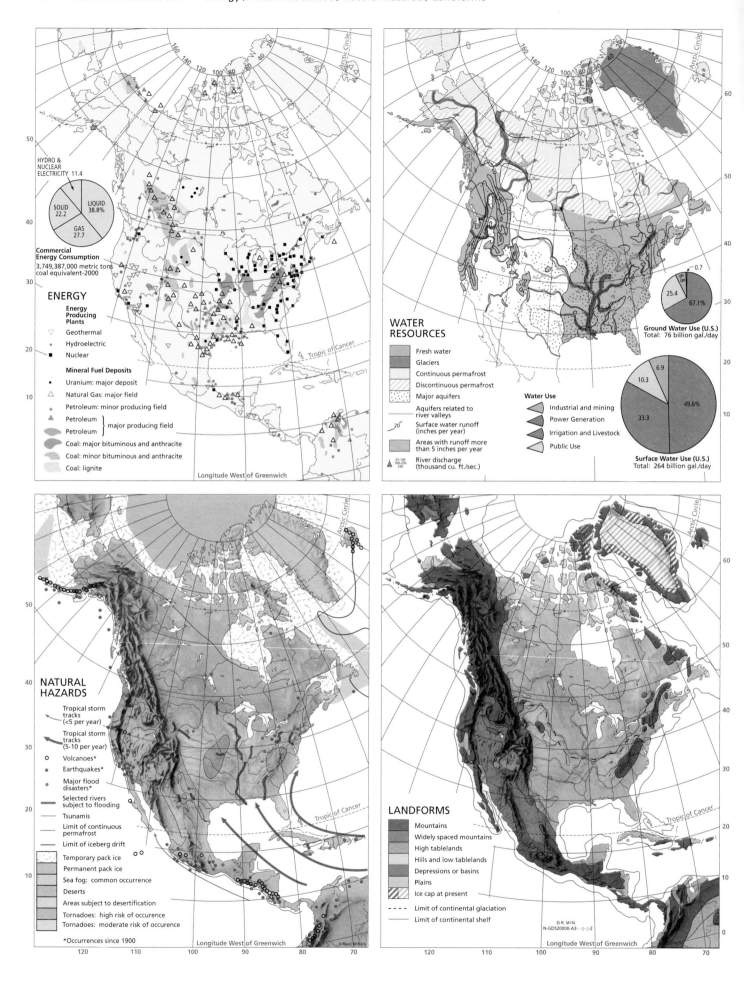

ENERGY

HYDRO & NUCLEAR ELECTRICITY 11.4
LIQUID 38.8%
SOLID 22.2
GAS 27.7

Commercial Energy Consumption 3,749,387,000 metric tons coal equivalent-2000

Energy Producing Plants
▽ Geothermal
• Hydroelectric
■ Nuclear

Mineral Fuel Deposits
• Uranium: major deposit
△ Natural Gas: major field
• Petroleum: minor producing field
▲ Petroleum } major producing field
Petroleum
Coal: major bituminous and anthracite
Coal: minor bituminous and anthracite
Coal: lignite

Longitude West of Greenwich

WATER RESOURCES

Fresh water
Glaciers
Continuous permafrost
Discontinuous permafrost
Major aquifers

Aquifers related to river valleys
—20— Surface water runoff (inches per year)
Areas with runoff more than 5 inches per year
River discharge (thousand cu. ft./sec.)

Ground Water Use (U.S.)
Total: 76 billion gal./day
0.7
6.9
25.4
67.1%

Water Use
◁ Industrial and mining
◁ Power Generation
◁ Irrigation and Livestock
◁ Public Use

Surface Water Use (U.S.)
Total: 264 billion gal./day
6.9
10.3
33.3
49.6%

NATURAL HAZARDS

← Tropical storm tracks (<5 per year)
← Tropical storm tracks (5-10 per year)
○ Volcanoes*
• Earthquakes*
• Major flood disasters*
— Selected rivers subject to flooding
Tsunamis
Limit of continuous permafrost
Limit of iceberg drift
Temporary pack ice
Permanent pack ice
Sea fog: common occurrence
Deserts
Areas subject to desertification
Tornadoes: high risk of occurence
Tornadoes: moderate risk of occurence

*Occurrences since 1900

Longitude West of Greenwich

LANDFORMS

Mountains
Widely spaced mountains
High tablelands
Hills and low tablelands
Depressions or basins
Plains
Ice cap at present

- - - Limit of continental glaciation
— Limit of continental shelf

© R. McN.
N-GDS20000-A3- -2-2-2

Longitude West of Greenwich

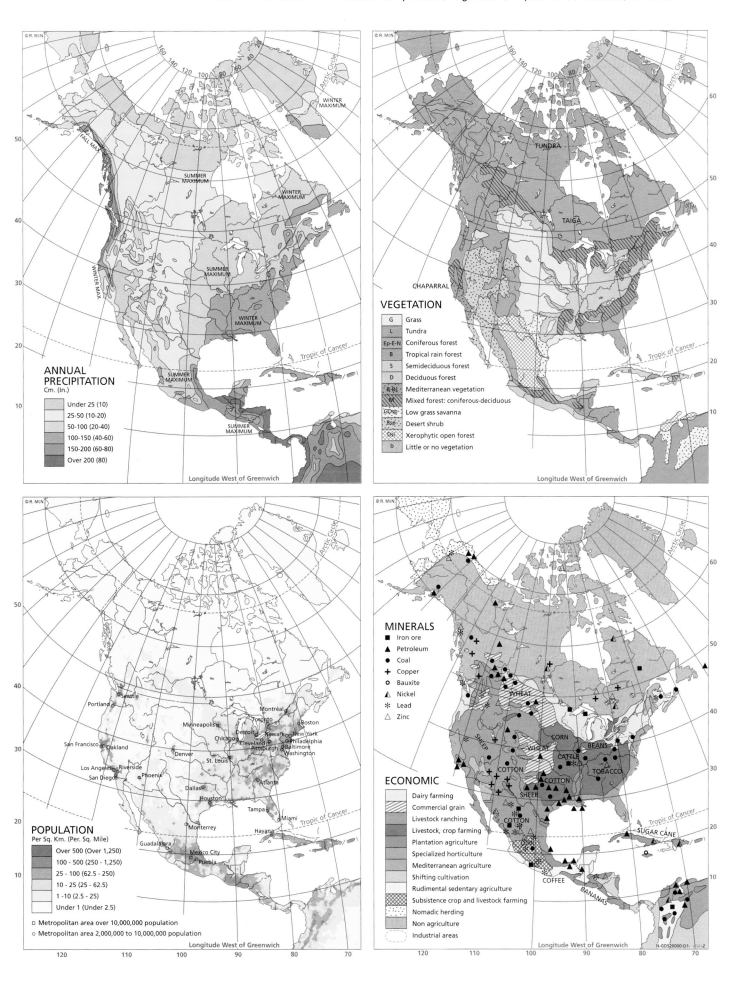

ANNUAL PRECIPITATION
Cm. (In.)

- Under 25 (10)
- 25-50 (10-20)
- 50-100 (20-40)
- 100-150 (40-60)
- 150-200 (60-80)
- Over 200 (80)

Longitude West of Greenwich

WINTER MAXIMUM
FALL MAX.
SUMMER MAXIMUM
WINTER MAXIMUM
WINTER MAX.
SUMMER MAXIMUM
SUMMER MAXIMUM
WINTER MAXIMUM
SUMMER MAXIMUM
SUMMER MAXIMUM

Tropic of Cancer

VEGETATION

G	Grass
L	Tundra
Ep-E-N	Coniferous forest
B	Tropical rain forest
S	Semideciduous forest
D	Deciduous forest
R-Bs	Mediterranean vegetation
M	Mixed forest: coniferous-deciduous
GDsp	Low grass savanna
Bsp	Desert shrub
Dsi	Xerophytic open forest
b	Little or no vegetation

TUNDRA
TAIGA
CHAPARRAL

Tropic of Cancer

Longitude West of Greenwich

POPULATION
Per Sq. Km. (Per. Sq. Mile)

- Over 500 (Over 1,250)
- 100 - 500 (250 - 1,250)
- 25 - 100 (62.5 - 250)
- 10 - 25 (25 - 62.5)
- 1 - 10 (2.5 - 25)
- Under 1 (Under 2.5)

□ Metropolitan area over 10,000,000 population
○ Metropolitan area 2,000,000 to 10,000,000 population

Seattle
Portland
San Francisco
Oakland
Los Angeles Riverside
San Diego
Phoenix
Denver
Dallas
Houston
Minneapolis
Chicago
Detroit
Cleveland
Pittsburgh
St. Louis
Atlanta
Tampa
Miami
Monterrey
Havana
Guadalajara
Mexico City
Puebla
Montréal
Toronto
Boston
Newark
New York
Philadelphia
Baltimore
Washington

Tropic of Cancer

Longitude West of Greenwich

MINERALS

- ■ Iron ore
- ▲ Petroleum
- ● Coal
- + Copper
- ○ Bauxite
- ▲ Nickel
- ✳ Lead
- △ Zinc

ECONOMIC

- Dairy farming
- Commercial grain
- Livestock ranching
- Livestock, crop farming
- Plantation agriculture
- Specialized horticulture
- Mediterranean agriculture
- Shifting cultivation
- Rudimental sedentary agriculture
- Subsistence crop and livestock farming
- Nomadic herding
- Non agriculture
- Industrial areas

WHEAT
SHEEP
WHEAT
CORN
BEANS
CATTLE
COTTON
TOBACCO
COTTON
SHEEP
COTTON
SUGAR CANE
COFFEE
BANANAS

Tropic of Cancer

Longitude West of Greenwich

N-GDS20000-D1-

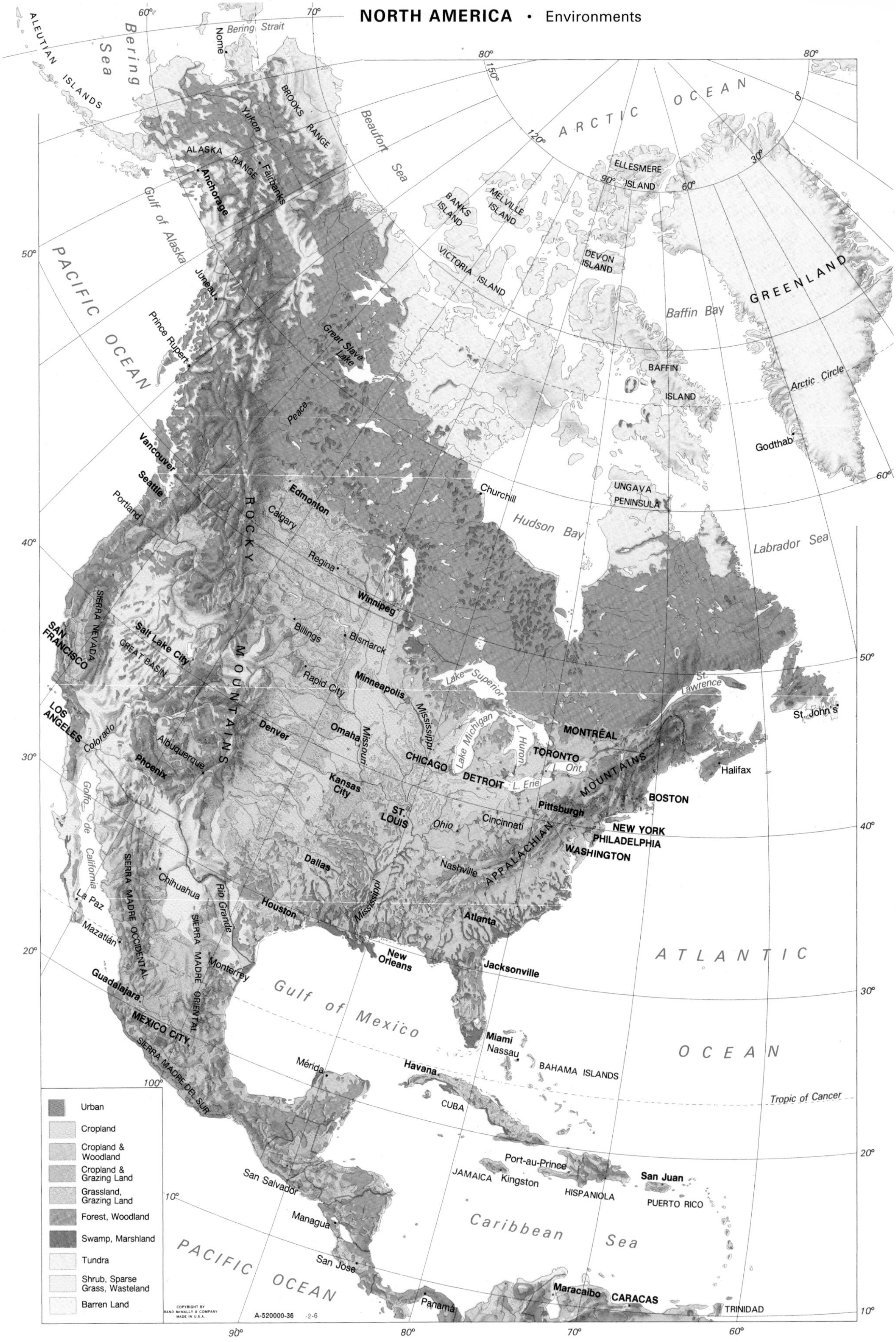

ALEUTIAN ISLANDS
Bering Strait
Nome
Bering Sea
ARCTIC OCEAN
ELLESMERE ISLAND
BROOKS RANGE
ALASKA RANGE
Anchorage
Fairbanks
Yukon
Beaufort Sea
BANKS ISLAND
MELVILLE ISLAND
DEVON ISLAND
GREENLAND
Gulf of Alaska
Juneau
VICTORIA ISLAND
Baffin Bay
PACIFIC OCEAN
Prince Rupert
BAFFIN ISLAND
Arctic Circle
Godthab
Vancouver
Seattle
Portland
Great Slave Lake
Peace
Churchill
Hudson Bay
UNGAVA PENINSULA
Labrador Sea
ROCKY MOUNTAINS
Edmonton
Calgary
Regina
Winnipeg
St. Lawrence
St. John's
SIERRA NEVADA
SAN FRANCISCO
GREAT BASIN
Salt Lake City
Billings
Bismarck
Rapid City
Lake Superior
L. Huron
Ont.
MONTRÉAL
TORONTO
L. Erie
Halifax
LOS ANGELES
Colorado
Denver
Omaha
Minneapolis
Missouri
Mississippi
Lake Michigan
CHICAGO
DETROIT
Pittsburgh
BOSTON
Albuquerque
Phoenix
Kansas City
ST. LOUIS
Ohio
Cincinnati
APPALACHIAN MOUNTAINS
NEW YORK
PHILADELPHIA
WASHINGTON
Golfo de California
SIERRA MADRE OCCIDENTAL
Chihuahua
Rio Grande
Dallas
Houston
Mississippi
Nashville
Atlanta
La Paz
Mazatlán
Monterrey
SIERRA MADRE ORIENTAL
New Orleans
Jacksonville
ATLANTIC OCEAN
Guadalajara
MEXICO CITY
SIERRA MADRE DEL SUR
Gulf of Mexico
Mérida
Havana
Miami
Nassau
BAHAMA ISLANDS
Tropic of Cancer
San Salvador
CUBA
San Jose
Managua
Port-au-Prince
JAMAICA
Kingston
HISPANIOLA
San Juan
PUERTO RICO
Caribbean Sea
PACIFIC OCEAN
San Jose
Panama
Maracaibo
CARACAS
TRINIDAD

Legend:
- Urban
- Cropland
- Cropland & Woodland
- Cropland & Grazing Land
- Grassland, Grazing Land
- Forest, Woodland
- Swamp, Marshland
- Tundra
- Shrub, Sparse Grass, Wasteland
- Barren Land

COPYRIGHT BY RAND MCNALLY & COMPANY MADE IN U.S.A.
A-520000-36 -2-6

Scale 1:36,000,000; one inch to 570 miles. Lambert Azimuthal Equal-Area Projection

0 100 200 400 600 800 Miles
0 150 300 600 900 1200 Kilometers

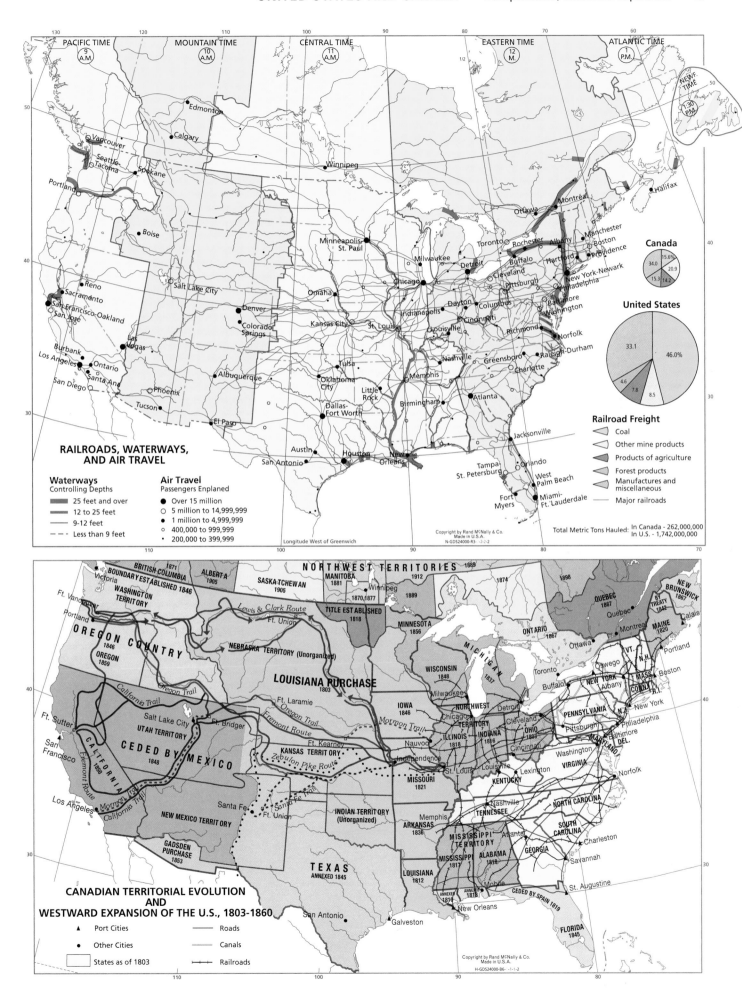

RAILROADS, WATERWAYS, AND AIR TRAVEL

Waterways
Controlling Depths
- 25 feet and over
- 12 to 25 feet
- 9-12 feet
- Less than 9 feet

Air Travel
Passengers Enplaned
- ● Over 15 million
- ○ 5 million to 14,999,999
- • 1 million to 4,999,999
- ○ 400,000 to 999,999
- · 200,000 to 399,999

Canada
34.0 15.6
20.9
15.3 14.2

United States
46.0%
33.1
4.6 7.8 8.5

Railroad Freight
- Coal
- Other mine products
- Products of agriculture
- Forest products
- Manufactures and miscellaneous
- Major railroads

Total Metric Tons Hauled: In Canada - 262,000,000
In U.S. - 1,742,000,000

Copyright by Rand McNally & Co.
Made in U.S.A.
N-GDS24000-R3- -2-2-2

Longitude West of Greenwich

CANADIAN TERRITORIAL EVOLUTION AND WESTWARD EXPANSION OF THE U.S., 1803-1860

- ▲ Port Cities
- ● Other Cities
- ☐ States as of 1803
- ——— Roads
- ——— Canals
- ┼┼┼ Railroads

Copyright by Rand McNally & Co.
Made in U.S.A.
H-GDS24000-B6- -1-1-2

PHYSIOGRAPHIC DIVISIONS

1 Pacific Mountain System
2 Intermontane Plateaus
3 Rocky Mountain System
4 Interior Plains
5 Ozark-Ouachita Highlands
6 Gulf-Atlantic Plain
7 Appalachian Highlands
8 Laurentian Upland (Canadian Shield)
9 Hudson Bay Lowland

Scale 1:12 000 000; One inch to 190 miles. POLYCONIC PROJECTION

0 25 50 75 100 200 300 400 500 Miles

0 50 100 200 400 600 800 Kilometers

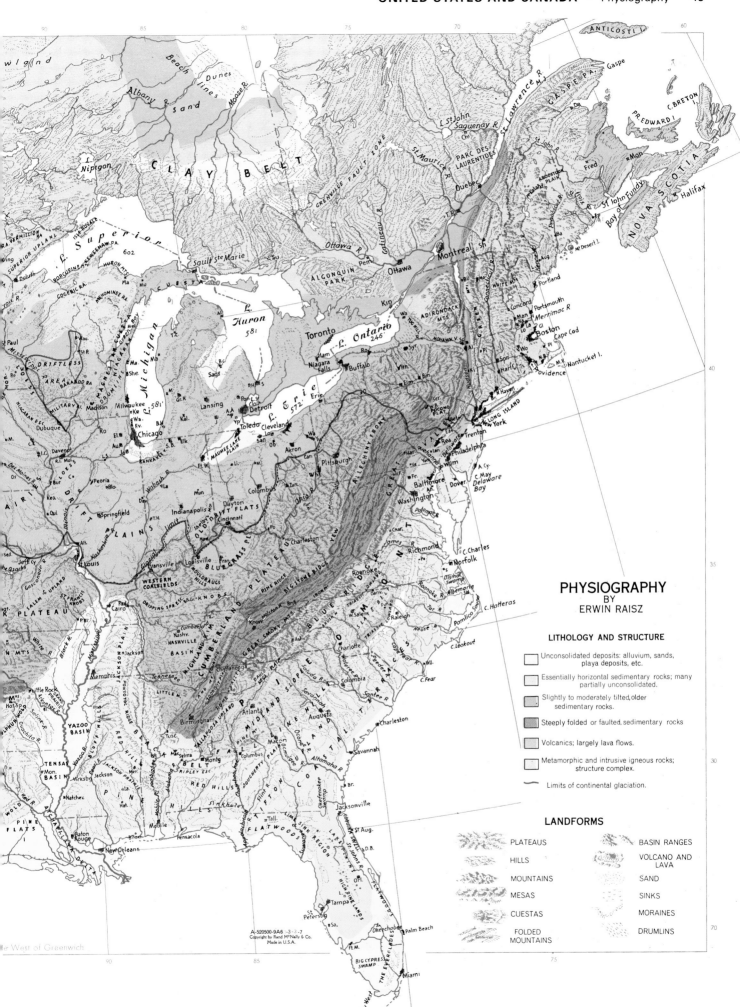

PHYSIOGRAPHY
BY
ERWIN RAISZ

LITHOLOGY AND STRUCTURE

Unconsolidated deposits: alluvium, sands, playa deposits, etc.

Essentially horizontal sedimentary rocks; many partially unconsolidated.

Slightly to moderately tilted, older sedimentary rocks.

Steeply folded or faulted, sedimentary rocks

Volcanics; largely lava flows.

Metamorphic and intrusive igneous rocks; structure complex.

Limits of continental glaciation.

LANDFORMS

PLATEAUS   BASIN RANGES

HILLS   VOLCANO AND LAVA

MOUNTAINS   SAND

MESAS   SINKS

CUESTAS   MORAINES

FOLDED MOUNTAINS   DRUMLINS

A-520500-9A6 -3-3-7
Copyright by Rand McNally & Co.
Made in U.S.A.

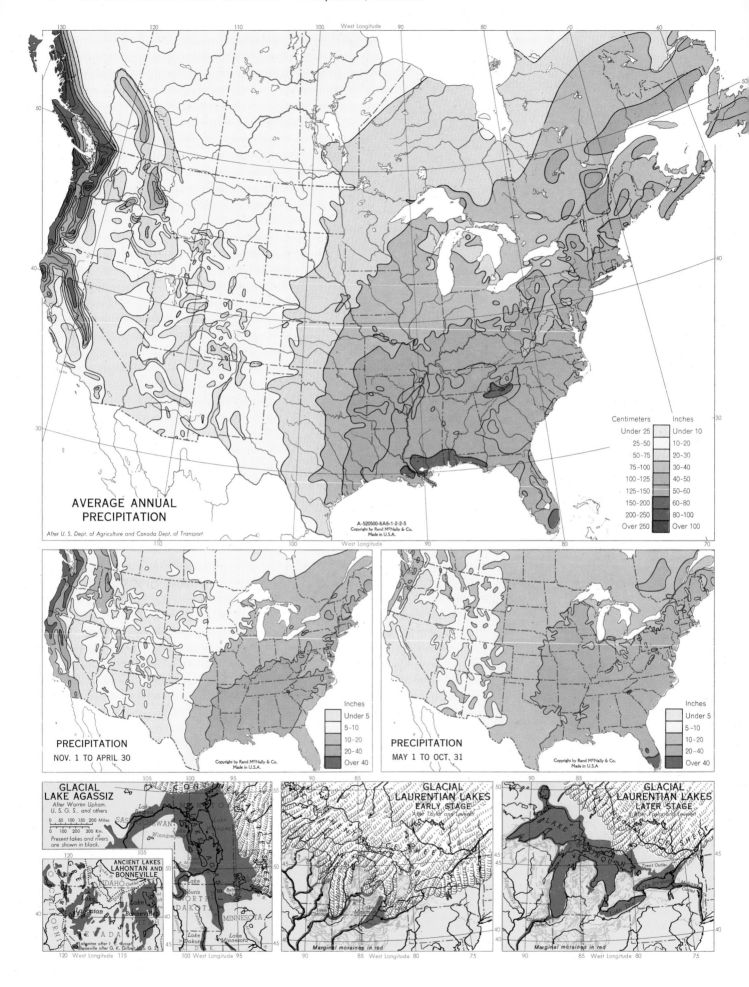

AVERAGE ANNUAL
PRECIPITATION

After U. S. Dept. of Agriculture and Canada Dept. of Transport

A-520500-6A6-1-2-2-5
Copyright by Rand McNally & Co.
Made in U.S.A.

Centimeters	Inches
Under 25	Under 10
25-50	10-20
50-75	20-30
75-100	30-40
100-125	40-50
125-150	50-60
150-200	60-80
200-250	80-100
Over 250	Over 100

PRECIPITATION

NOV. 1 TO APRIL 30

Copyright by Rand McNally & Co.
Made in U.S.A.

Inches
Under 5
5-10
10-20
20-40
Over 40

PRECIPITATION

MAY 1 TO OCT. 31

Copyright by Rand McNally & Co.
Made in U.S.A.

Inches
Under 5
5-10
10-20
20-40
Over 40

GLACIAL
LAKE AGASSIZ

After Warren Upham,
U. S. G. S., and others

0 50 100 150 200 Miles
0 100 200 300 Km.
Present lakes and rivers
are shown in black.

ANCIENT LAKES
LAHONTAN AND
BONNEVILLE

Lahontan after I. C. Russell
Bonneville after G. K. Gilbert, U. S. G. S.

GLACIAL
LAURENTIAN LAKES
EARLY STAGE

After Taylor and Leverett

Marginal moraines in red

GLACIAL
LAURENTIAN LAKES
LATER STAGE

After Taylor and Leverett

Marginal moraines in red

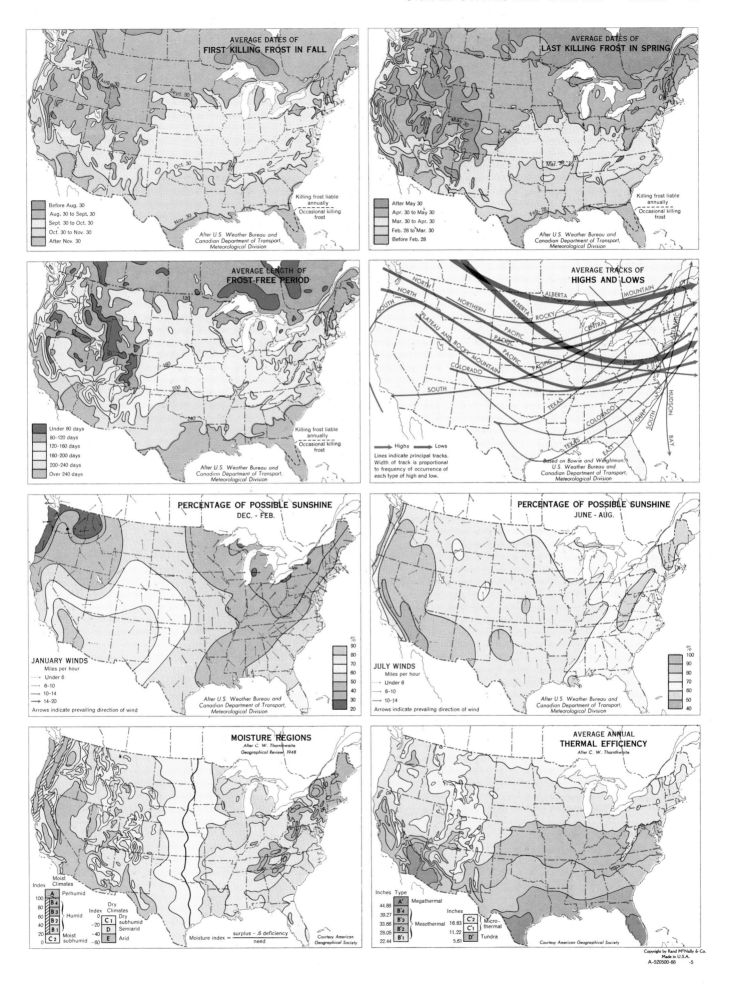

AVERAGE DATES OF
FIRST KILLING FROST IN FALL

Before Aug. 30
Aug. 30 to Sept. 30
Sept. 30 to Oct. 30
Oct. 30 to Nov. 30
After Nov. 30

Killing frost liable
annually
Occasional killing
frost

After U.S. Weather Bureau and
Canadian Department of Transport,
Meteorological Division

AVERAGE DATES OF
LAST KILLING FROST IN SPRING

After May 30
Apr. 30 to May 30
Mar. 30 to Apr. 30
Feb. 28 to Mar. 30
Before Feb. 28

Killing frost liable
annually
Occasional killing
frost

After U.S. Weather Bureau and
Canadian Department of Transport,
Meteorological Division

AVERAGE LENGTH OF
FROST-FREE PERIOD

Under 80 days
80–120 days
120–160 days
160–200 days
200–240 days
Over 240 days

Killing frost liable
annually
Occasional killing
frost

After U.S. Weather Bureau and
Canadian Department of Transport,
Meteorological Division

AVERAGE TRACKS OF
HIGHS AND LOWS

Highs   Lows

Lines indicate principal tracks.
Width of track is proportional
to frequency of occurrence of
each type of high and low.

Based on Bowie and Weightman.
U.S. Weather Bureau and
Canadian Department of Transport,
Meteorological Division

PERCENTAGE OF POSSIBLE SUNSHINE
DEC. - FEB.

JANUARY WINDS
Miles per hour
Under 6
6–10
10–14
14–20
Arrows indicate prevailing direction of wind

%
90
80
70
60
50
40
30
20

After U.S. Weather Bureau and
Canadian Department of Transport,
Meteorological Division

PERCENTAGE OF POSSIBLE SUNSHINE
JUNE - AUG.

JULY WINDS
Miles per hour
Under 6
6–10
10–14
Arrows indicate prevailing direction of wind

%
100
90
80
70
60
50
40

After U.S. Weather Bureau and
Canadian Department of Transport,
Meteorological Division

MOISTURE REGIONS
After C. W. Thornthwaite
Geographical Review, 1948

Index
100
80
60
40
20
0

Moist
Climates
A   Perhumid
B4
B3   Humid
B2
B1
C2   Moist
       subhumid

Index
0
-20
-40
-60

Dry
Climates
C1   Dry
D   subhumid
       Semiarid
E   Arid

Moisture index = surplus − .6 deficiency / need

Courtesy American
Geographical Society

AVERAGE ANNUAL
THERMAL EFFICIENCY
After C. W. Thornthwaite

Inches   Type
44.88   A'   Megathermal
39.27   B'4
33.66   B'3   Mesothermal
28.05   B'2
22.44   B'1

Inches
16.83   C'2   Micro-
11.22   C'1   thermal
 5.61   D'   Tundra

Courtesy American Geographical Society

Copyright by Rand M°Nally & Co.
Made in U.S.A.
A-520500-66   -5

48

KEY TO CLASSIFICATION

B - Broadleaf evergreen
D - Broadleaf deciduous
E - Needleleaf evergreen
G - Grass
L - Herbaceous plants other than grass
N - Needleleaf deciduous
O - Woody plants without leaves
b - Vegetation largely or entirely absent
l - Low; maximum height of trees 30 feet, maximum
     height of herbaceous plants 1½ feet
m - Medium height; maximum height of trees 30-75 feet,
     maximum height of herbaceous plants 1½ -6 feet
p - Growth singly or in groups or patches
s - Shrubform, minimum height 3 feet
z - Dwarf shrubform, maximum height 3 feet

   The various formulas are used to designate types of
vegetation on this map. Each formula constitutes a short
description of the chief characteristics of a vegetation.
The classification is based on whether plants are woody
or herbaceous, and if woody, whether they are broadleaf
or needleleaf and evergreen or deciduous. The small
letters are added to give more detail to the description.
   All capital letters other than G and L imply trees, un-
less accompanied by s or z. The small letters refer to
the capital letter immediately preceding them. Thus,
GlDsp means that the vegetation consists of low grass
(Gl) and of patches of broadleaf deciduous shrubs
(Dsp); EDp represents needleleaf evergreen trees (E) with
patches of broadleaf deciduous trees (Dp).

B Broadleaf evergreen trees

1 Mangrove

Bs Broadleaf evergreen, shrubform

2 Ceanothus-manzanita-chamise

Bz Broadleaf evergreen, dwarf shrubform

3 Greasewood
4 Sagebrush
5 Sage-sagebrush

Bsz Broadleaf evergreen, shubform and
     dwarf shrubform

6 Creosote bush
7 Lechuquilla-sotol

Bzp Broadleaf evergreen, dwarf shrubform,
     in patches

8 Shadscale

BzGm Broadleaf evergreen, dwarf shrubform
     Grass, medium height

9 Sandsage-sandgrass

Scale 1:14 000 000; One inch to

0 25 50 75 100   200   300   400   500 Miles

0 50 100   400   600   800 Kilometers

NATURAL VEGETATION

BY A. W. KÜCHLER

Based on "A Physiognomic Classification of Vegetation"
Annals of the Assoc. of American Geographers, Vol. 39, September, 1949

D Broadleaf deciduous trees

10 Aspen-oak
11 Beech-maple
12 Beech-tulip tree-maple-basswood
13 Cottonwood-willow
14 Maple-basswood
15 Oak
16 Oak-ash-maple
17 Oak-hickory
18 Oak-tulip tree

DB Broadleaf deciduous trees
Broadleaf evergreen trees

19 Oak-madrone

DE Broadleaf deciduous trees
Needleleaf evergreen trees

20 Maple-yellow birch-hemlock-pine
21 Oak-Douglas fir
22 Oak-pine
23 Maple-beech-hemlock

D / Gmp Broadleaf deciduous trees
Grass, medium height, in patches

24 Aspen-needle grass-wheat grass
25 Oak-hickory-bluestem

DN Broadleaf deciduous trees
Needleleaf deciduous trees

26 Bay trees-bald cypress
27 Tupelo-gum-bald cypress

E Needleleaf evergreen trees

28 Douglas fir
29 Douglas fir-redwood
30 Hemlock-arbor vitae
31 Hemlock-arbor vitae-Douglas fir
32 Hemlock-arbor vitae-fir
33 Hemlock-spruce
34 Pine
35 Pine-juniper
36 Pine-spruce
37 Spruce-fir

Esp Needleleaf evergreen, shrubform, in patches

38 Juniper

EDp Needleleaf evergreen trees
Broadleaf deciduous trees, in patches

39 Douglas fir-pine-aspen
40 Pine-spruce-birch
41 Spruce-aspen
42 Spruce-fir-aspen
43 Spruce-poplar-birch

EN Needleleaf evergreen trees
Needleleaf deciduous trees

44 Hemlock-arbor vitae-Douglas fir-larch
45 Pine-bald cypress
46 Pine-spruce-larch
47 Spruce-larch

Gl Grass, low

48 Grama grass
49 Grama grass-buffalo grass
50 Grama grass-needle grass
51 Needle grass-blue grass
52 Wheat grass
53 Wheat grass-blue grass

Gm Grass, medium height

54 Bluestem
55 Broom grass-water grass
56 Marsh grass
57 Saw grass

Gml Grass, medium and low height

58 Bluestem-bunch grass
59 Needle grass-wheat grass

Gl / Dsp Grass, low
Broadleaf deciduous, shrubform, in patches

60 Bunch grass-oak

Gm / Dsp Grass, medium height
Broadleaf deciduous, shrubform, in patches

61 Mesquite grass-mesquite

L Herbaceous plants other than grass

62 Lichens, etc.

LEp Herbaceous plants other than grass
Needleleaf evergreen trees, in patches

63 Lichens-spruce

LEp / Np Herbaceous plants other than grass
Needleleaf evergreen trees, in patches
Needleleaf deciduous trees, in patches

64 Lichens-spruce-larch

N Needleleaf deciduous trees

65 Bald cypress

Op Woody plants without leaves, in patches

66 Palo verde-cacti-ocotillo

b Vegetation largely or entirely absent

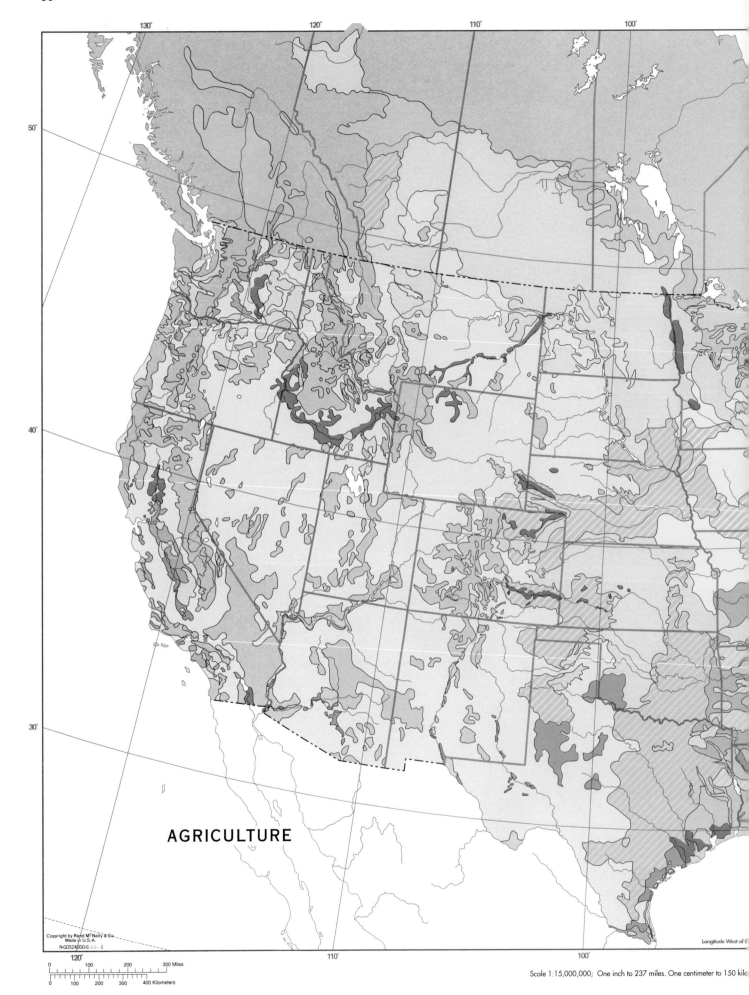

130° 120° 110° 100°

50°

40°

30°

AGRICULTURE

120°

110° 100°

Longitude West of G

0   100   200   300 Miles
0   100   200   300   400 Kilometers

Scale 1:15,000,000; One inch to 237 miles. One centimeter to 150 kilo

Dairying

Fruits and Vegetables

Wheat, Barley, and Oilseeds

Cash Corn and Soybeans

Tobacco

Cotton

Livestock and Feed Grains: Beef

Livestock and Feed Grains: Hogs

Livestock and Feed Grains: Poultry

Livestock and Feed Grains: Mixed

Specialty Crops (Peanuts, Potatoes, Rice, Sugar)

Western Livestock Ranching

Western Feedlots

Agriculture and Forestry

Non-Agricultural Areas

S CONIC PROJECTION

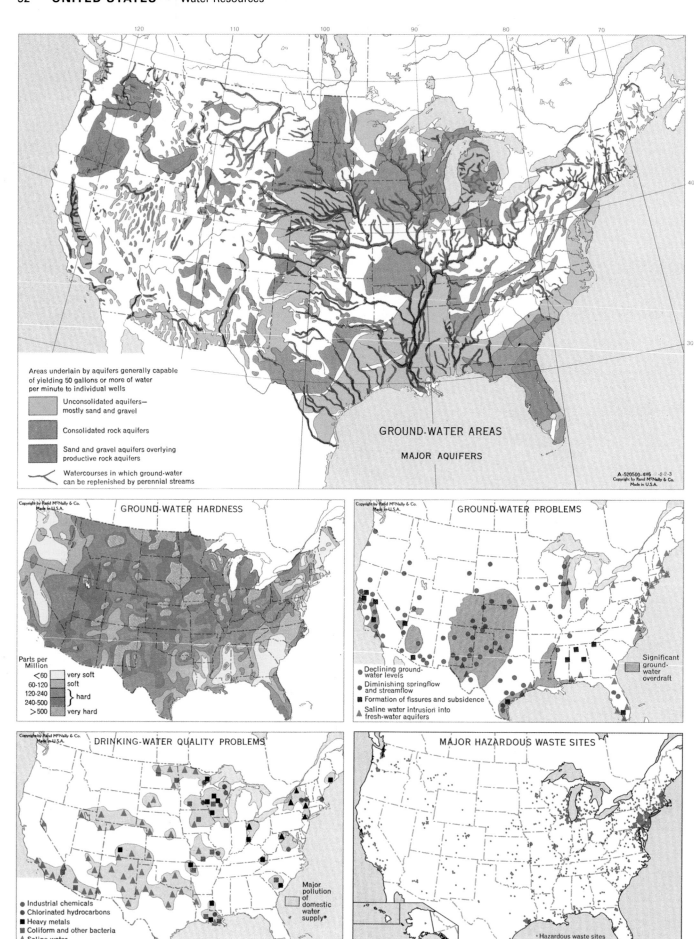

Areas underlain by aquifers generally capable
of yielding 50 gallons or more of water
per minute to individual wells

Unconsolidated aquifers—
mostly sand and gravel

Consolidated rock aquifers

Sand and gravel aquifers overlying
productive rock aquifers

Watercourses in which ground-water
can be replenished by perennial streams

GROUND-WATER AREAS

MAJOR AQUIFERS

A-520500-4H6
Copyright by Rand McNally & Co.
Made in U.S.A.

GROUND-WATER HARDNESS

Parts per
Million
<60   very soft
60-120   soft
120-240 } hard
240-500
>500   very hard

GROUND-WATER PROBLEMS

• Declining ground-
  water levels
• Diminishing springflow
  and streamflow
■ Formation of fissures and subsidence
▲ Saline water intrusion into
  fresh-water aquifers

Significant
ground-
water
overdraft

DRINKING-WATER QUALITY PROBLEMS

● Industrial chemicals
● Chlorinated hydrocarbons
■ Heavy metals
■ Coliform and other bacteria
▲ Saline water
▲ Municipal and industrial wastes

Major
pollution
of
domestic
water
supply*

*Potential and existing

MAJOR HAZARDOUS WASTE SITES

• Hazardous waste sites

©RMCN.

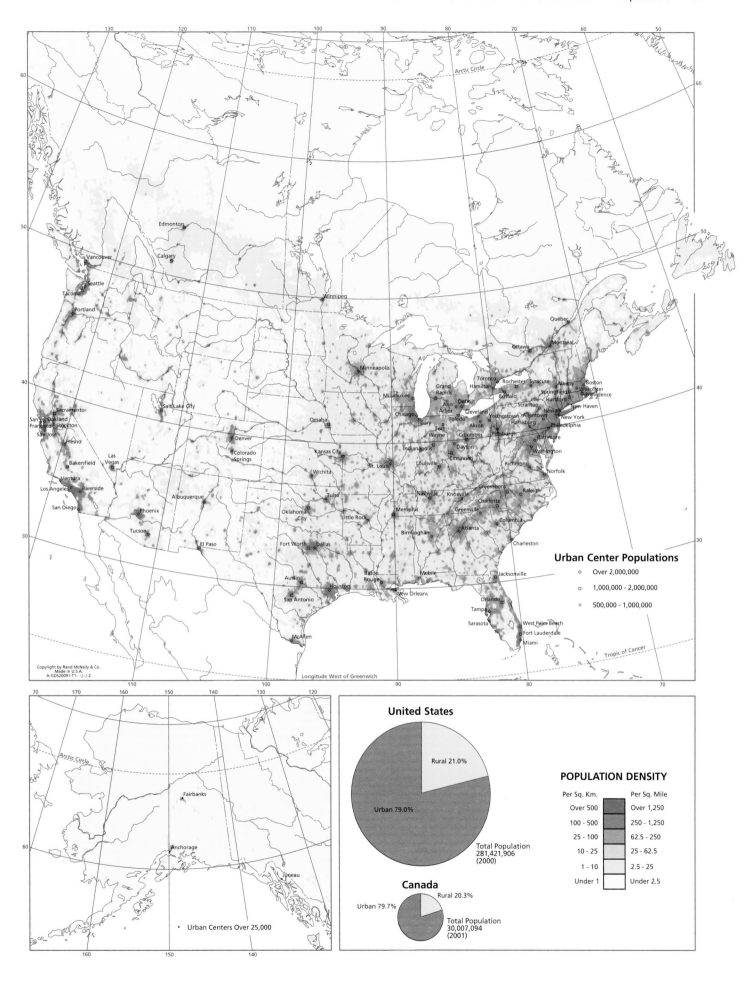

Urban Center Populations

◇ Over 2,000,000

▫ 1,000,000 - 2,000,000

○ 500,000 - 1,000,000

Copyright by Rand McNally & Co.
Made in U.S.A.
A-GDS20091-T1- -2-2-2

Longitude West of Greenwich

· Urban Centers Over 25,000

United States

Rural 21.0%

Urban 79.0%

Total Population
281,421,906
(2000)

Canada

Rural 20.3%

Urban 79.7%

Total Population
30,007,094
(2001)

POPULATION DENSITY

Per Sq. Km.	Per Sq. Mile
Over 500	Over 1,250
100 - 500	250 - 1,250
25 - 100	62.5 - 250
10 - 25	25 - 62.5
1 - 10	2.5 - 25
Under 1	Under 2.5

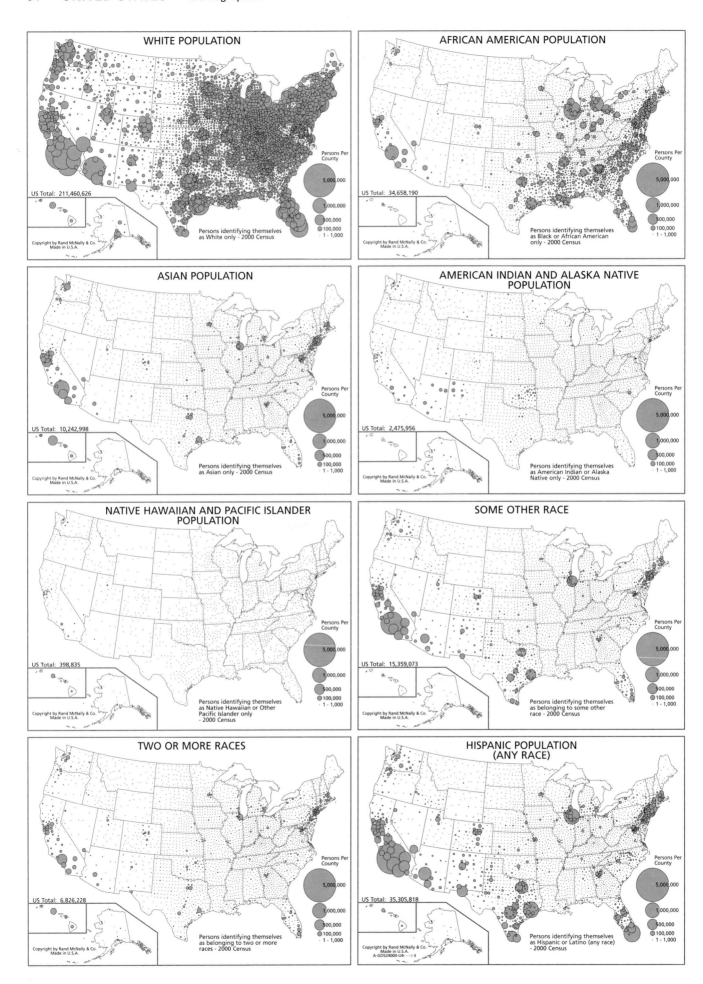

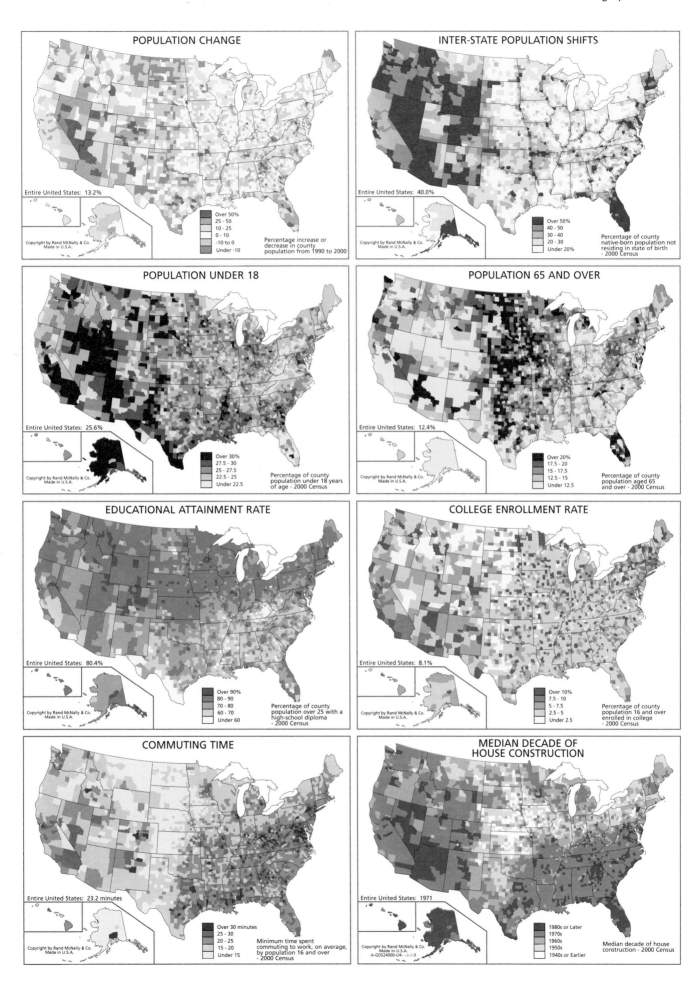

POPULATION CHANGE

Entire United States: 13.2%

Over 50%
25 - 50
10 - 25
0 - 10
-10 to 0
Under -10

Percentage increase or decrease in county population from 1990 to 2000

Copyright by Rand McNally & Co. Made in U.S.A.

INTER-STATE POPULATION SHIFTS

Entire United States: 40.0%

Over 50%
40 - 50
30 - 40
20 - 30
Under 20%

Percentage of county native-born population not residing in state of birth - 2000 Census

Copyright by Rand McNally & Co. Made in U.S.A.

POPULATION UNDER 18

Entire United States: 25.6%

Over 30%
27.5 - 30
25 - 27.5
22.5 - 25
Under 22.5

Percentage of county population under 18 years of age - 2000 Census

Copyright by Rand McNally & Co. Made in U.S.A.

POPULATION 65 AND OVER

Entire United States: 12.4%

Over 20%
17.5 - 20
15 - 17.5
12.5 - 15
Under 12.5

Percentage of county population aged 65 and over - 2000 Census

Copyright by Rand McNally & Co. Made in U.S.A.

EDUCATIONAL ATTAINMENT RATE

Entire United States: 80.4%

Over 90%
80 - 90
70 - 80
60 - 70
Under 60

Percentage of county population over 25 with a high-school diploma - 2000 Census

Copyright by Rand McNally & Co. Made in U.S.A.

COLLEGE ENROLLMENT RATE

Entire United States: 8.1%

Over 10%
7.5 - 10
5 - 7.5
2.5 - 5
Under 2.5

Percentage of county population 16 and over enrolled in college - 2000 Census

Copyright by Rand McNally & Co. Made in U.S.A.

COMMUTING TIME

Entire United States: 23.2 minutes

Over 30 minutes
25 - 30
20 - 25
15 - 20
Under 15

Minimum time spent commuting to work, on average, by population 16 and over - 2000 Census

Copyright by Rand McNally & Co. Made in U.S.A.

MEDIAN DECADE OF HOUSE CONSTRUCTION

Entire United States: 1971

1980s or Later
1970s
1960s
1950s
1940s or Earlier

Median decade of house construction - 2000 Census

Copyright by Rand McNally & Co. Made in U.S.A.
A-GD524000-U4- -3-3-3

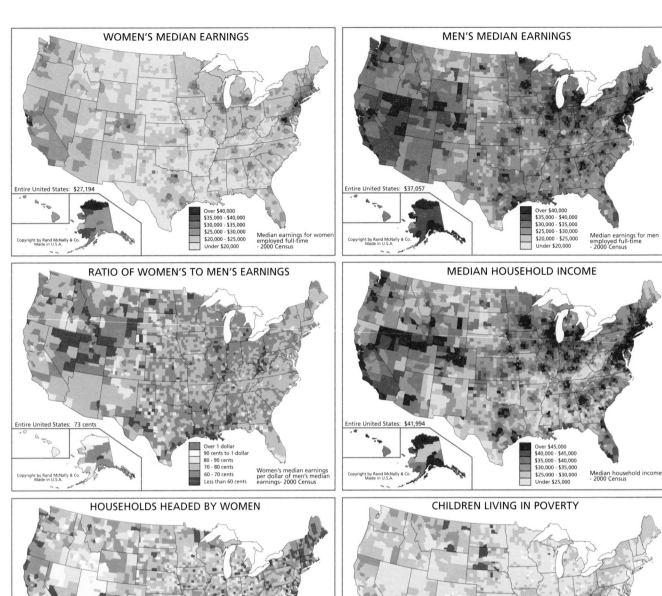

WOMEN'S MEDIAN EARNINGS

Entire United States: $27,194

Over $40,000
$35,000 - $40,000
$30,000 - $35,000
$25,000 - $30,000
$20,000 - $25,000
Under $20,000

Median earnings for women employed full-time - 2000 Census

Copyright by Rand McNally & Co. Made in U.S.A.

MEN'S MEDIAN EARNINGS

Entire United States: $37,057

Over $40,000
$35,000 - $40,000
$30,000 - $35,000
$25,000 - $30,000
$20,000 - $25,000
Under $20,000

Median earnings for men employed full-time - 2000 Census

Copyright by Rand McNally & Co. Made in U.S.A.

RATIO OF WOMEN'S TO MEN'S EARNINGS

Entire United States: 73 cents

Over 1 dollar
90 cents to 1 dollar
80 - 90 cents
70 - 80 cents
60 - 70 cents
Less than 60 cents

Women's median earnings per dollar of men's median earnings- 2000 Census

Copyright by Rand McNally & Co. Made in U.S.A.

MEDIAN HOUSEHOLD INCOME

Entire United States: $41,994

Over $45,000
$40,000 - $45,000
$35,000 - $40,000
$30,000 - $35,000
$25,000 - $30,000
Under $25,000

Median household income - 2000 Census

Copyright by Rand McNally & Co. Made in U.S.A.

HOUSEHOLDS HEADED BY WOMEN

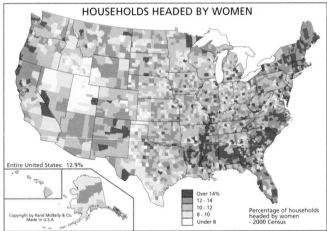

Entire United States: 12.9%

Over 14%
12 - 14
10 - 12
8 - 10
Under 8

Percentage of households headed by women - 2000 Census

Copyright by Rand McNally & Co. Made in U.S.A.

CHILDREN LIVING IN POVERTY

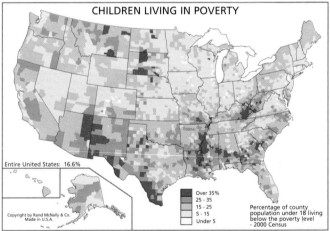

Entire United States: 16.6%

Over 35%
25 - 35
15 - 25
5 - 15
Under 5

Percentage of county population under 18 living below the poverty level - 2000 Census

Copyright by Rand McNally & Co. Made in U.S.A.

UNEMPLOYMENT RATE

Entire United States: 5.8%

Over 10%
8 - 10
6 - 8
4 - 6
Under 4

Percentage of civilian labor force unemployed - 2000 Census

Copyright by Rand McNally & Co. Made in U.S.A.

NON-ENGLISH SPEAKING HOUSEHOLDS

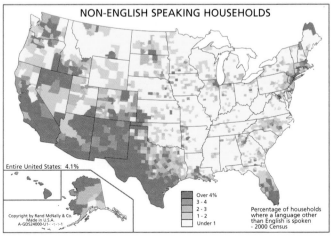

Entire United States: 4.1%

Over 4%
3 - 4
2 - 3
1 - 2
Under 1

Percentage of households where a language other than English is spoken - 2000 Census

Copyright by Rand McNally & Co. Made in U.S.A.
A-GDS24000-U1- -1-1-1

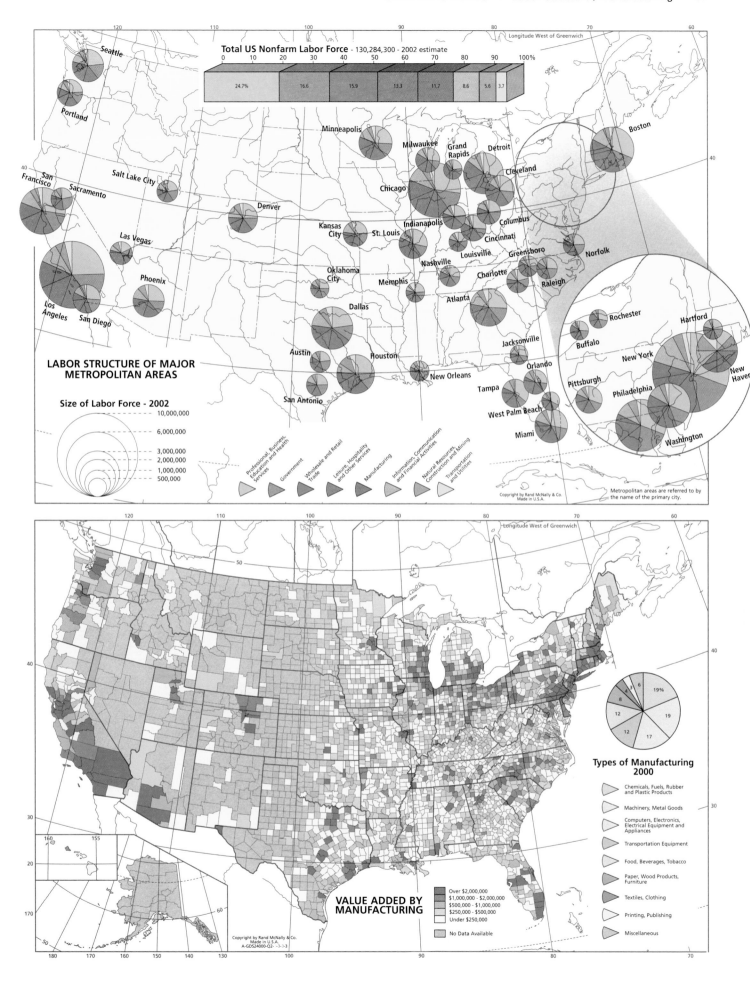

Total US Nonfarm Labor Force - 130,284,300 - 2002 estimate

24.7%	16.6	15.9	13.3	11.7	8.6	5.6	3.7

LABOR STRUCTURE OF MAJOR METROPOLITAN AREAS

Size of Labor Force - 2002

- 10,000,000
- 6,000,000
- 3,000,000
- 2,000,000
- 1,000,000
- 500,000

Professional, Business, Education and Health Services

Government

Wholesale and Retail Trade

Leisure, Hospitality and Other Services

Manufacturing

Information Communication and Financial Activities

Natural Resources, Construction and Mining

Transportation and Utilities

Copyright by Rand McNally & Co.
Made in U.S.A.

Metropolitan areas are referred to by the name of the primary city.

Longitude West of Greenwich

VALUE ADDED BY MANUFACTURING

- Over $2,000,000
- $1,000,000 - $2,000,000
- $500,000 - $1,000,000
- $250,000 - $500,000
- Under $250,000
- No Data Available

Copyright by Rand McNally & Co.
Made in U.S.A.
A-GDS24000-Q2- -3-3-3

Types of Manufacturing 2000

- Chemicals, Fuels, Rubber and Plastic Products
- Machinery, Metal Goods
- Computers, Electronics, Electrical Equipment and Appliances
- Transportation Equipment
- Food, Beverages, Tobacco
- Paper, Wood Products, Furniture
- Textiles, Clothing
- Printing, Publishing
- Miscellaneous

Scale 1:12,000,000.
One inch to 190 miles.
One centimeter to 120 kilometers.
Albers Conic Projection

0   50 100   200   300   400 Miles

0 50 100 150 200   300   400   500   600 Kilometers

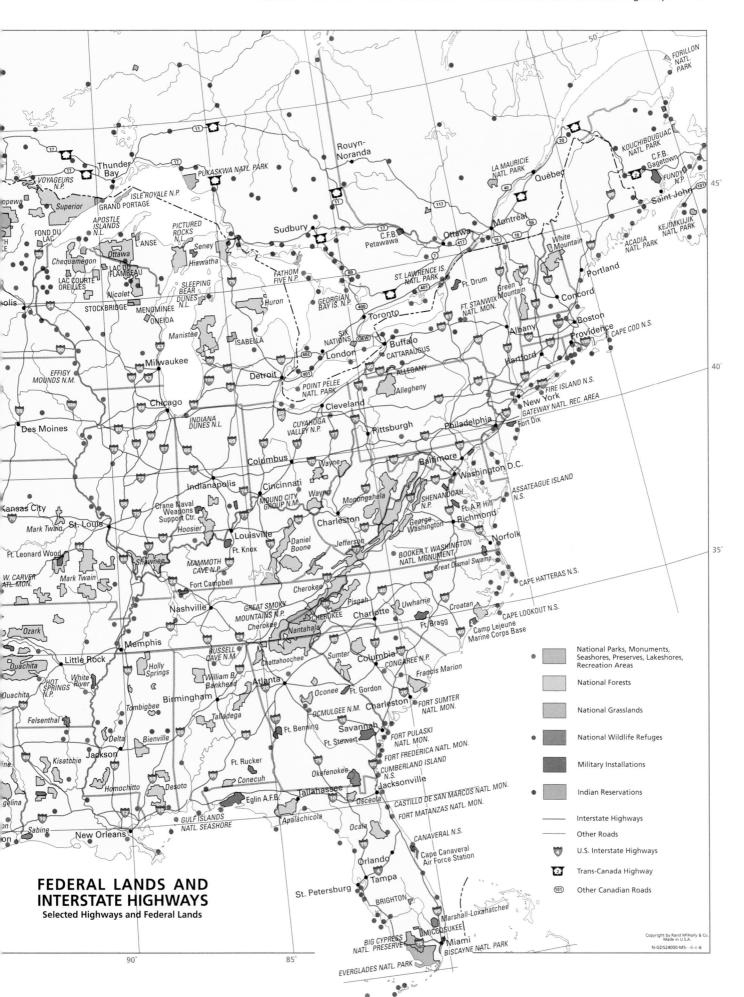

FEDERAL LANDS AND INTERSTATE HIGHWAYS
Selected Highways and Federal Lands

▨	National Parks, Monuments, Seashores, Preserves, Lakeshores, Recreation Areas
▨	National Forests
▨	National Grasslands
●	National Wildlife Refuges
▨	Military Installations
▨	Indian Reservations
──	Interstate Highways
──	Other Roads
🛡	U.S. Interstate Highways
🛡	Trans-Canada Highway
101	Other Canadian Roads

Copyright by Rand McNally & Co.
Made in U.S.A.

N-GD524000-MS- -6-6-6

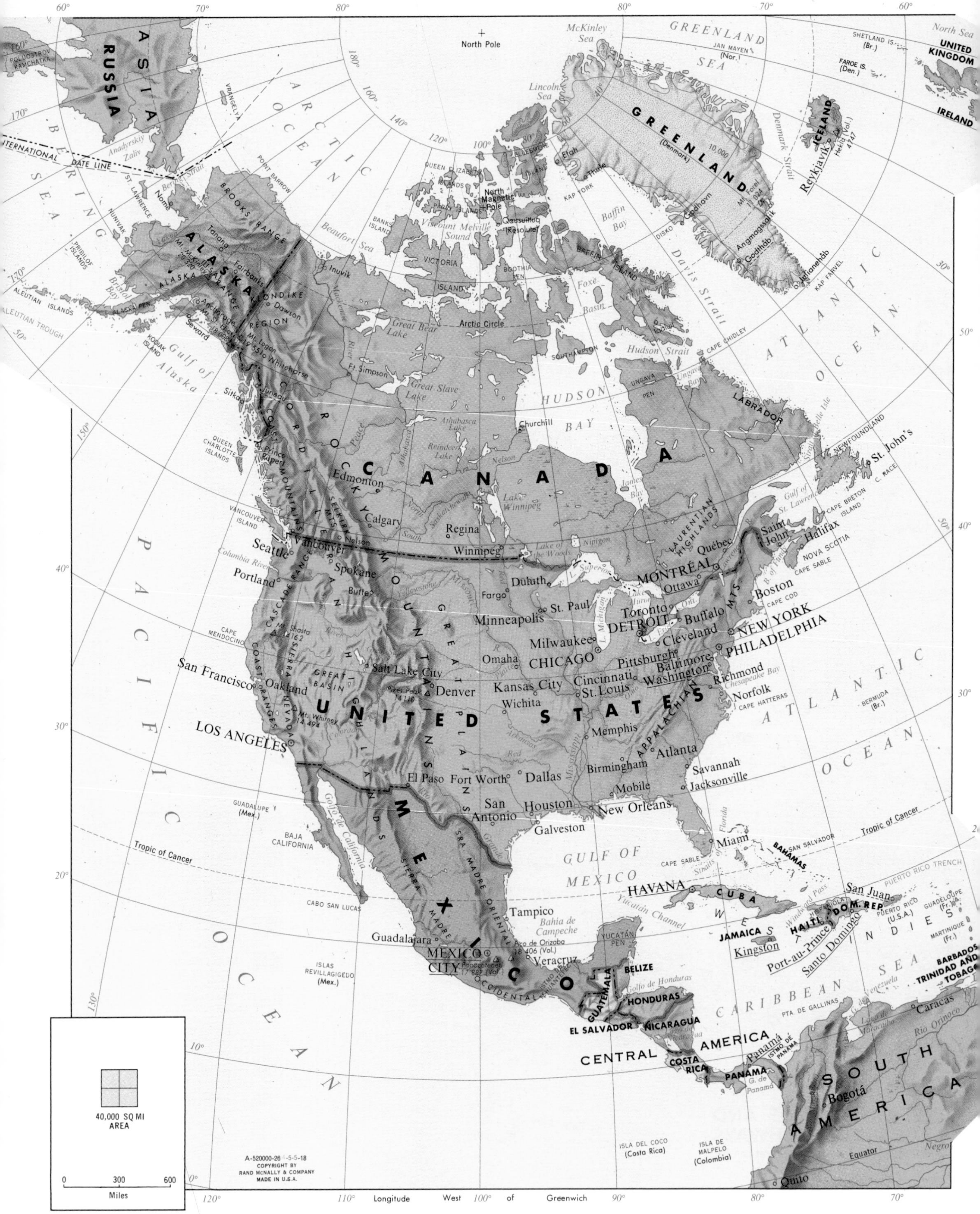

North Pole
ARCTIC OCEAN
GREENLAND SEA
GREENLAND (Denmark)
McKinley Sea
Lincoln Sea
JAN MAYEN (Nor.)
SHETLAND IS. (Br.)
FAROE IS. (Den.)
North Sea
UNITED KINGDOM
IRELAND
ICELAND
Reykjavík
Denmark Strait
ASIA
RUSSIA
POLUOSTROV KAMCHATKA
VRANGELYA
Anadyrskiy Zaliv
INTERNATIONAL DATE LINE
BERING SEA
Bering Strait
ST. LAWRENCE
NUNIVAK
PRIBILOF ISLANDS
ALEUTIAN ISLANDS
ALEUTIAN TROUGH
Bristol Bay
KODIAK ISLAND
POINT BARROW
BROOKS RANGE
ALASKA
ALASKA RANGE
Nome
Tanana
Fairbanks
Mt. McKinley 20,320
Anchorage
Seward
KLONDIKE REGION
Dawson
Whitehorse
Mt. Logan 19,551
Gulf of Alaska
Sitka
Juneau
Beaufort Sea
Inuvik
Mackenzie
BANKS ISLAND
VICTORIA ISLAND
QUEEN ELIZABETH ISLANDS
North Magnetic Pole
Viscount Melville Sound
Qausuittuq (Resolute)
BOOTHIA PEN.
ELLESMERE ISLAND
Etah
Thule
KAP YORK
Baffin Bay
DISKO
Godhavn
Angmagssalik
Godthåb
Julianehåb
KAP FARVEL
Mt. Forel 11,024
Hekla (Vol.) 4747
BAFFIN ISLAND
Davis Strait
Foxe Basin
Iqaluit
Arctic Circle
Great Bear Lake
Ft. Simpson
Great Slave Lake
SOUTHAMPTON
Hudson Strait
CAPE CHIDLEY
UNGAVA PEN.
Ungava Bay
HUDSON BAY
LABRADOR
Churchill
Athabasca Lake
Reindeer Lake
Nelson
Lake Winnipeg
James Bay
NEWFOUNDLAND
St. John's
C. RACE
Gulf of St. Lawrence
CAPE BRETON ISLAND
Halifax
NOVA SCOTIA
CAPE SABLE
Saint John
LAURENTIAN HIGHLANDS
CANADA
CORDILLERA
ROCKY MOUNTAINS
COAST MOUNTAINS
QUEEN CHARLOTTE ISLANDS
Prince Rupert
VANCOUVER ISLAND
Edmonton
Calgary
Regina
Winnipeg
Lake of the Woods
Nipigon
L. Superior
Vancouver
Seattle
Nelson
SELKIRK MTS.
Spokane
Portland
Columbia River
Butte
Yellowstone
Missouri
CASCADE RANGE
Mt. Shasta 14,162
CAPE MENDOCINO
COAST RANGES
SIERRA NEVADA
Mt. Whitney 14,494
San Francisco
Oakland
LOS ANGELES
GREAT BASIN
Salt Lake City
Pikes Peak 14,110
Denver
GREAT PLAINS
Fargo
Duluth
St. Paul
Minneapolis
L. Michigan
Lake Huron
Toronto
DETROIT
Buffalo
L. Erie
Cleveland
Milwaukee
CHICAGO
Omaha
Platte
Kansas City
Wichita
St. Louis
Cincinnati
Pittsburgh
Baltimore
Washington
Richmond
Chesapeake Bay
Norfolk
CAPE HATTERAS
Ottawa
MONTRÉAL
Québec
Boston
CAPE COD
NEW YORK
PHILADELPHIA
APPALACHIAN MTS.
UNITED STATES
Memphis
Arkansas
Red
Atlanta
Birmingham
Savannah
Jacksonville
Mobile
New Orleans
Mississippi
El Paso
Fort Worth
Dallas
San Antonio
Houston
Galveston
Rio Grande
BERMUDA (Br.)
ATLANTIC OCEAN
PACIFIC OCEAN
GUADALUPE (Mex.)
BAJA CALIFORNIA
Golfo de California
Tropic of Cancer
SIERRA MADRE OCCIDENTAL
SRA. MADRE ORIENTAL
MEXICO
CABO SAN LUCAS
Tampico
Bahía de Campeche
Guadalajara
MEXICO CITY
Pico de Orizaba 18,406 (Vol.)
Popocatépetl 17,887 (Vol.)
Veracruz
ISLAS REVILLAGIGEDO (Mex.)
YUCATÁN PEN.
GULF OF MEXICO
Florida
Miami
CAPE SABLE
Straits of Florida
BAHAMAS
SAN SALVADOR
HAVANA
CUBA
Yucatán Channel
Windward Pass
JAMAICA
Kingston
HAITI
Port-au-Prince
DOM. REP.
Santo Domingo
San Juan
PUERTO RICO (U.S.A.)
PUERTO RICO TRENCH
GUADELOUPE (Fr.)
MARTINIQUE (Fr.)
BARBADOS
TRINIDAD AND TOBAGO
WEST INDIES
CARIBBEAN SEA
BELIZE
Golfo de Honduras
GUATEMALA
HONDURAS
EL SALVADOR
NICARAGUA
COSTA RICA
PANAMA
Panamá
ISTMO DE PANAMÁ
G. de Panamá
CENTRAL AMERICA
PTA DE GALLINAS
Lago de Maracaibo
Caracas
Río Orinoco
Bogotá
Quito
Equator
SOUTH AMERICA
ISLA DEL COCO (Costa Rica)
ISLA DE MALPELO (Colombia)
40,000 SQ MI AREA
0 300 600 Miles
A-520000-26-5-5-18
COPYRIGHT BY RAND McNALLY & COMPANY
MADE IN U.S.A.
Longitude West of Greenwich

0 200 400 600 800 1000 Miles
0 400 800 1200 1600 Kilometers

Scale 1:40 000 000; one inch to 630 miles. Lambert's Azimuthal Equal Area Projection
Elevations and depressions are given in feet

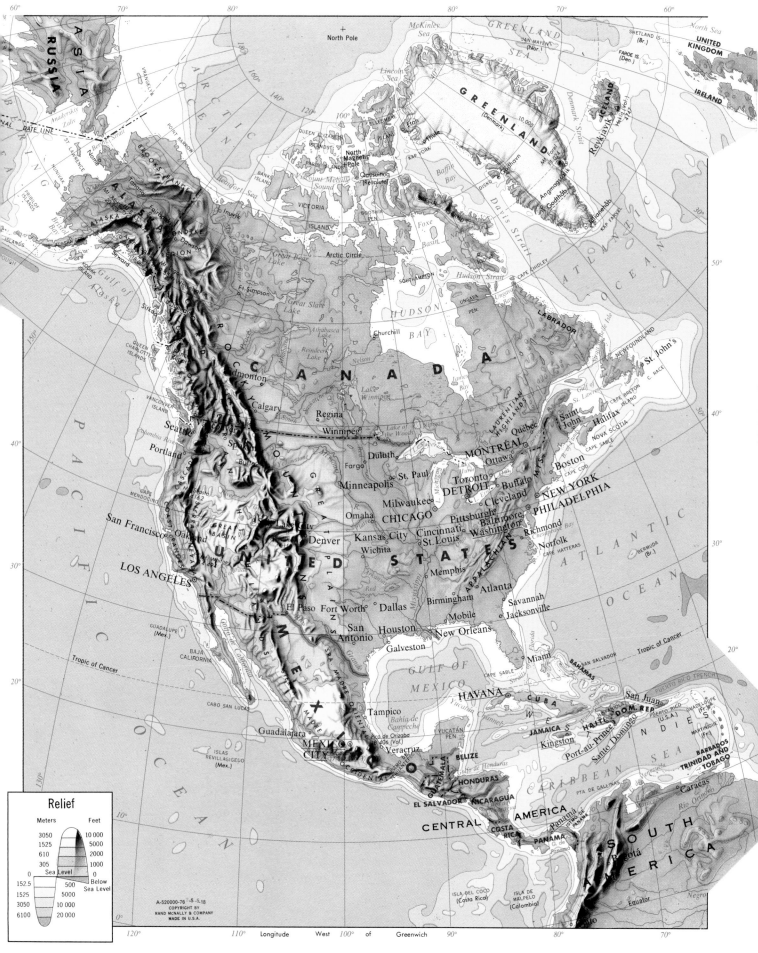

Relief

Meters   Feet

3050   10 000
1525   5000
610   2000
305   1000
0   Sea Level   0
152.5   500
0   0
Below   Below
Sea Level   Sea Level
1525   5000
3050   10 000
6100   20 000

A-520000-76 -5 -5 18
COPYRIGHT BY
RAND McNALLY & COMPANY
MADE IN U.S.A.

0   200   400   600   800   1000 Miles
0   400   800   1200   1600 Kilometers

Scale 1:40 000 000; one inch to 630 miles. Lambert's Azimuthal Equal Area Projection
Elevations and depressions are given in feet

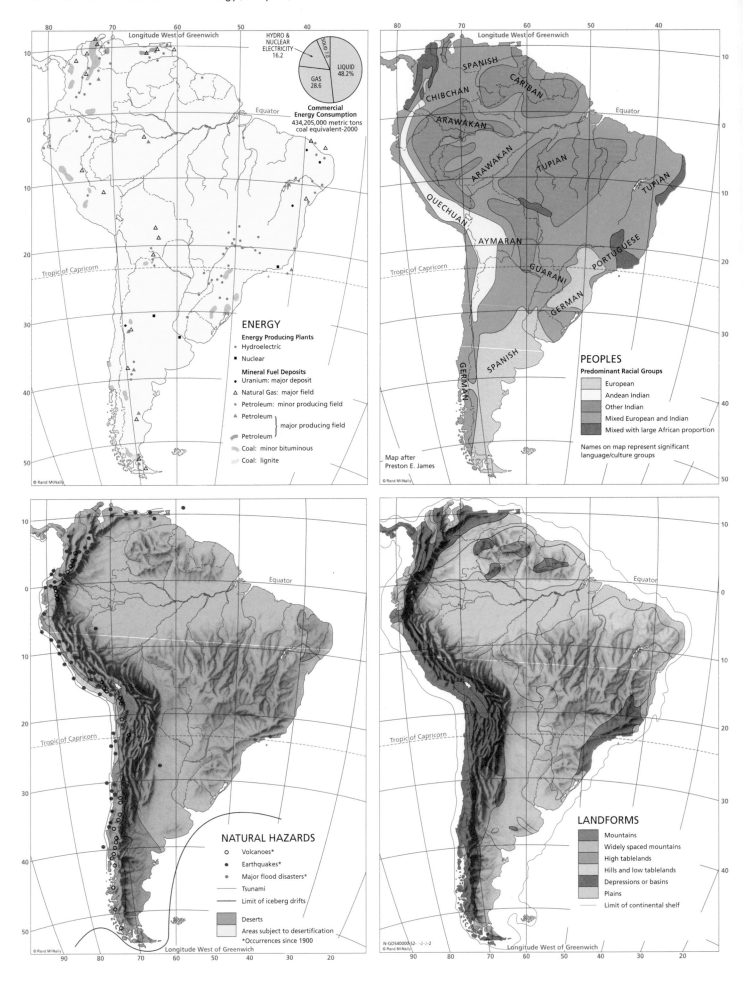

ENERGY

Energy Producing Plants
- Hydroelectric
- Nuclear

Mineral Fuel Deposits
- Uranium: major deposit
- △ Natural Gas: major field
- Petroleum: minor producing field
- ▲ Petroleum
- Petroleum } major producing field
- Coal: minor bituminous
- Coal: lignite

HYDRO & NUCLEAR ELECTRICITY 16.2
SOLID 7.0
LIQUID 48.2%
GAS 28.6

Commercial Energy Consumption
434,205,000 metric tons coal equivalent-2000

© Rand McNally

PEOPLES

Predominant Racial Groups
- European
- Andean Indian
- Other Indian
- Mixed European and Indian
- Mixed with large African proportion

Names on map represent significant language/culture groups

Map after Preston E. James

© Rand McNally

SPANISH
CHIBCHAN
CARIBAN
ARAWAKAN
ARAWAKAN
TUPIAN
TUPIAN
QUECHUAN
AYMARAN
GUARANI
PORTUGUESE
GERMAN
GERMAN
SPANISH

NATURAL HAZARDS

- ○ Volcanoes*
- ● Earthquakes*
- ● Major flood disasters*
- Tsunami
- Limit of iceberg drifts
- Deserts
- Areas subject to desertification

*Occurrences since 1900

© Rand McNally

LANDFORMS

- Mountains
- Widely spaced mountains
- High tablelands
- Hills and low tablelands
- Depressions or basins
- Plains
- Limit of continental shelf

N-GDS40000AS2- -2-2-2
© Rand McNally

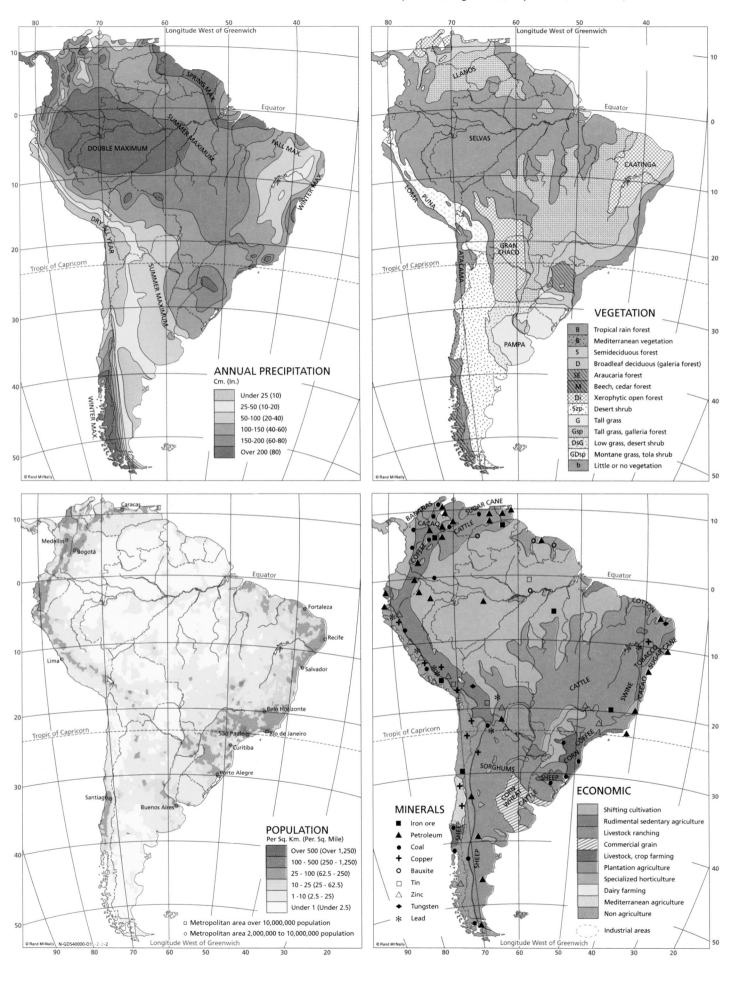

ANNUAL PRECIPITATION
Cm. (In.)

- Under 25 (10)
- 25-50 (10-20)
- 50-100 (20-40)
- 100-150 (40-60)
- 150-200 (60-80)
- Over 200 (80)

VEGETATION

B	Tropical rain forest
B'	Mediterranean vegetation
S	Semideciduous forest
D	Broadleaf deciduous (galeria forest)
SE	Araucaria forest
M	Beech, cedar forest
Di	Xerophytic open forest
Szp	Desert shrub
G	Tall grass
Gsp	Tall grass, galleria forest
DsG	Low grass, desert shrub
GDsp	Montane grass, tola shrub
b	Little or no vegetation

POPULATION
Per Sq. Km. (Per. Sq. Mile)

- Over 500 (Over 1,250)
- 100 - 500 (250 - 1,250)
- 25 - 100 (62.5 - 250)
- 10 - 25 (25 - 62.5)
- 1 - 10 (2.5 - 25)
- Under 1 (Under 2.5)

□ Metropolitan area over 10,000,000 population
○ Metropolitan area 2,000,000 to 10,000,000 population

MINERALS

- ■ Iron ore
- ▲ Petroleum
- ● Coal
- ✛ Copper
- ○ Bauxite
- □ Tin
- △ Zinc
- ◆ Tungsten
- ✳ Lead

ECONOMIC

- Shifting cultivation
- Rudimental sedentary agriculture
- Livestock ranching
- Commercial grain
- Livestock, crop farming
- Plantation agriculture
- Specialized horticulture
- Dairy farming
- Mediterranean agriculture
- Non agriculture
- Industrial areas

EUROPE LANGUAGES
BY
BOGDAN ZABORSKI

B-550000-1C6-1-1-1-4
COPYRIGHT BY
RAND MCNALLY & COMPANY
MADE IN U.S.A.

Scale 1:16,500,000; one inch to 260 miles Conic Projection

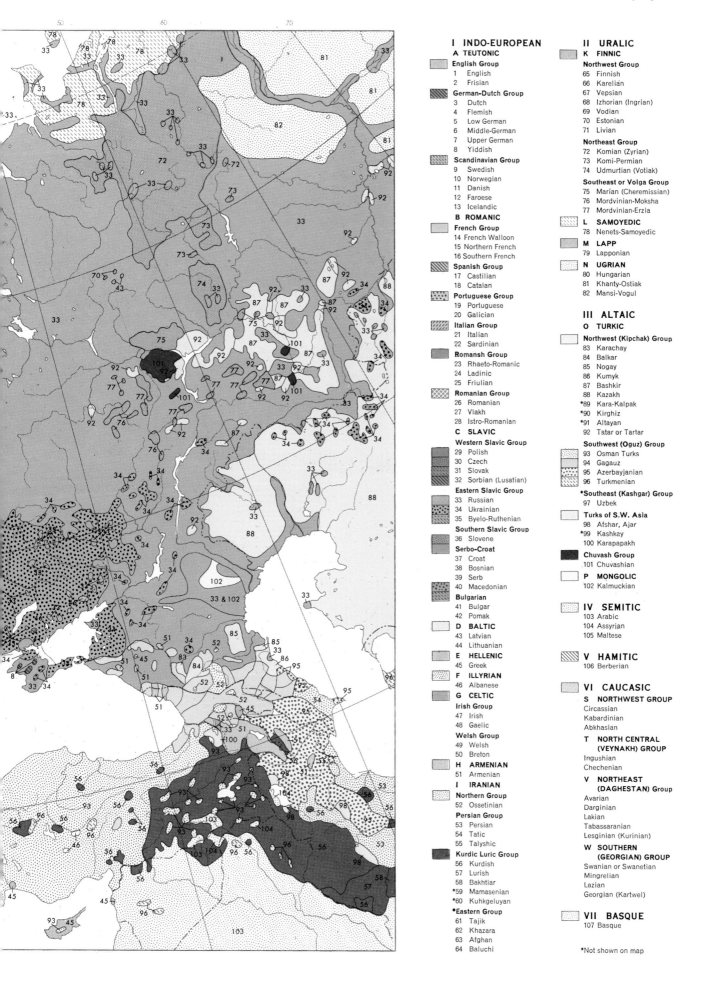

I INDO-EUROPEAN
A TEUTONIC
English Group
1 English
2 Frisian
German–Dutch Group
3 Dutch
4 Flemish
5 Low German
6 Middle-German
7 Upper German
8 Yiddish
Scandinavian Group
9 Swedish
10 Norwegian
11 Danish
12 Faroese
13 Icelandic
B ROMANIC
French Group
14 French Walloon
15 Northern French
16 Southern French
Spanish Group
17 Castilian
18 Catalan
Portuguese Group
19 Portuguese
20 Galician
Italian Group
21 Italian
22 Sardinian
Romansh Group
23 Rhaeto-Romanic
24 Ladinic
25 Friulian
Romanian Group
26 Romanian
27 Vlakh
28 Istro-Romanian
C SLAVIC
Western Slavic Group
29 Polish
30 Czech
31 Slovak
32 Sorbian (Lusatian)
Eastern Slavic Group
33 Russian
34 Ukrainian
35 Byelo-Ruthenian
Southern Slavic Group
36 Slovene
Serbo-Croat
37 Croat
38 Bosnian
39 Serb
40 Macedonian
Bulgarian
41 Bulgar
42 Pomak
D BALTIC
43 Latvian
44 Lithuanian
E HELLENIC
45 Greek
F ILLYRIAN
46 Albanese
G CELTIC
Irish Group
47 Irish
48 Gaelic
Welsh Group
49 Welsh
50 Breton
H ARMENIAN
51 Armenian
I IRANIAN
Northern Group
52 Ossetinian
Persian Group
53 Persian
54 Tatic
55 Talyshic
Kurdic Luric Group
56 Kurdish
57 Lurish
58 Bakhtiar
*59 Mamasenian
*60 Kuhgeluyan
*Eastern Group
61 Tajik
62 Khazara
63 Afghan
64 Baluchi

II URALIC
K FINNIC
Northwest Group
65 Finnish
66 Karelian
67 Vepsian
68 Izhorian (Ingrian)
69 Vodian
70 Estonian
71 Livian
Northeast Group
72 Komian (Zyrian)
73 Komi-Permian
74 Udmurtian (Votiak)
Southeast or Volga Group
75 Marian (Cheremissian)
76 Mordvinian-Moksha
77 Mordvinian-Erzia
L SAMOYEDIC
78 Nenets-Samoyedic
M LAPP
79 Lapponian
N UGRIAN
80 Hungarian
81 Khanty-Ostiak
82 Mansi-Vogul

III ALTAIC
O TURKIC
Northwest (Kipchak) Group
83 Karachay
84 Balkar
85 Nogay
86 Kumyk
87 Bashkir
88 Kazakh
*89 Kara-Kalpak
*90 Kirghiz
*91 Altayan
92 Tatar or Tartar
Southwest (Oguz) Group
93 Osman Turks
94 Gagauz
95 Azerbayjanian
96 Turkmenian
*Southeast (Kashgar) Group
97 Uzbek
Turks of S.W. Asia
98 Afshar, Ajar
*99 Kashkay
100 Karapapakh
Chuvash Group
101 Chuvashian
P MONGOLIC
102 Kalmuckian

IV SEMITIC
103 Arabic
104 Assyrian
105 Maltese

V HAMITIC
106 Berberian

VI CAUCASIC
S NORTHWEST GROUP
Circassian
Kabardinian
Abkhasian
**T NORTH CENTRAL
(VEYNAKH) GROUP**
Ingushian
Chechenian
**V NORTHEAST
(DAGHESTAN) Group**
Avarian
Darginian
Lakian
Tabassaranian
Lesginian (Kurinian)
**W SOUTHERN
(GEORGIAN) GROUP**
Swanian or Swanetian
Mingrelian
Lazian
Georgian (Kartwel)

VII BASQUE
107 Basque

*Not shown on map

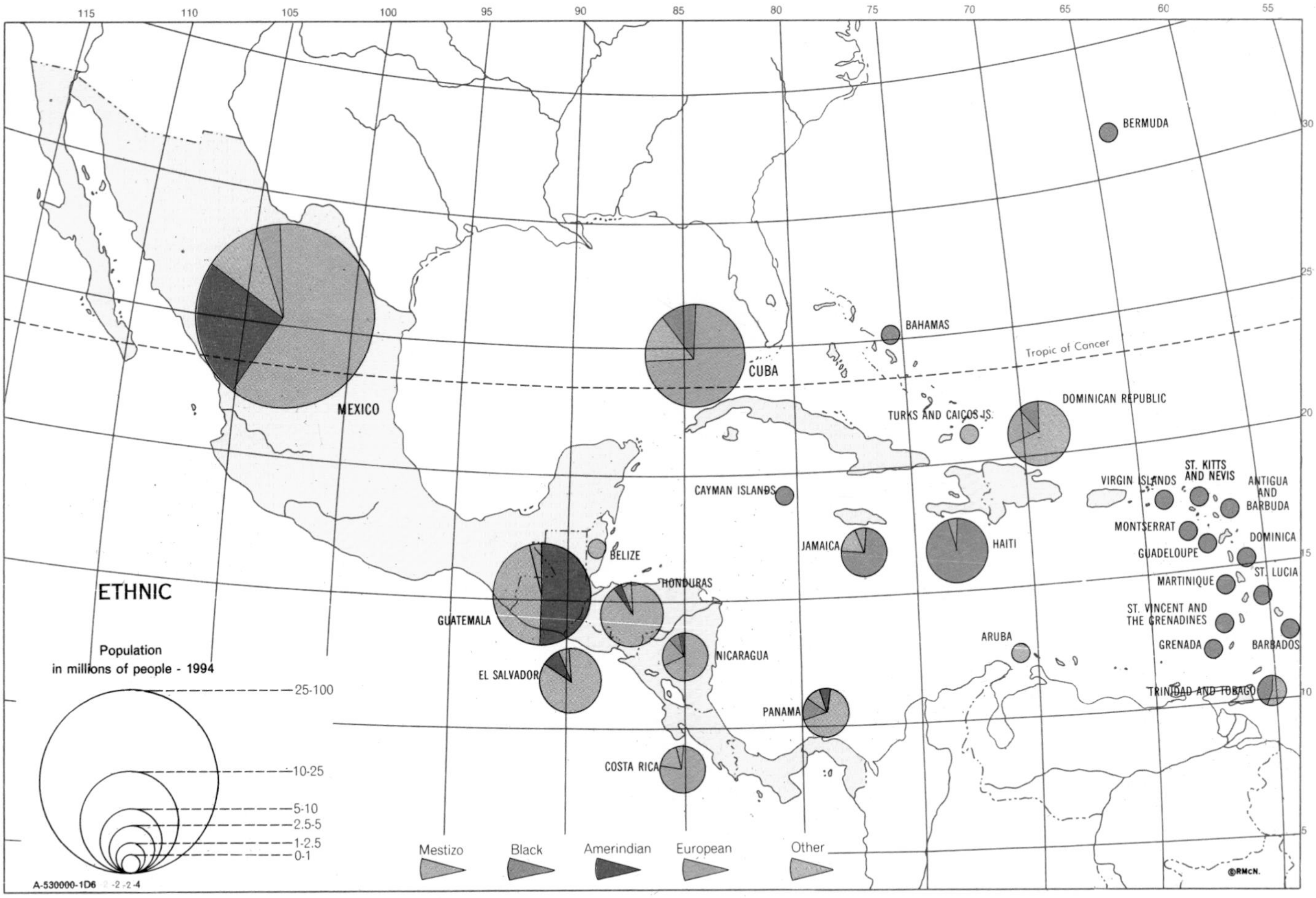

BERMUDA

CUBA

BAHAMAS

DOMINICAN REPUBLIC

TURKS AND CAICOS IS.

MEXICO

CAYMAN ISLANDS

VIRGIN ISLANDS

ST. KITTS AND NEVIS

ANTIGUA AND BARBUDA

MONTSERRAT

DOMINICA

JAMAICA

HAITI

GUADELOUPE

ST. LUCIA

BELIZE

MARTINIQUE

ETHNIC

HONDURAS

ST. VINCENT AND THE GRENADINES

GUATEMALA

Population
in millions of people - 1994

ARUBA

GRENADA

BARBADOS

EL SALVADOR

NICARAGUA

TRINIDAD AND TOBAGO

25-100

10-25

PANAMA

5-10

2.5-5

COSTA RICA

1-2.5

0-1

A-530000-1D6   -2 -2 -4

Mestizo   Black   Amerindian   European   Other

©RMcN.

U N I T E D   S T A T E S

BERMUDA
(Br.)

A T L A N T I C

· Hermosillo

O C E A N

M
E
X
I
C
O

· Monterrey

G U L F   O F   M E X I C O

B
A
H
A
M
A
S

Tropic of Cancer

· Tampico

Nassau ·

TURKS AND CAICOS
ISLANDS (Br.)

· Guadalajara

CUBA

DOMINICAN
REPUBLIC

PUERTO VIRGIN
RICO ISLANDS
(U.S.) (U.S.)(Br.)

ANGUILLA (Br.)

· Mérida

Havana ·

CAYMAN ISLANDS
(Br.)

HAITI

ANTIGUA AND
BARBUDA

· Mexico City

JAMAICA

Port-au-Prince ·

Santo Domingo ·

ST. KITTS AND NEVIS
MONTSERRAT (Br.)

· Oaxaca

BELIZE
Belmopan ·

Kingston ·

GUADELOUPE
(Fr.)

DOMINICA

C A R I B B E A N   S E A

MARTINIQUE(Fr.)

G
U
A
T
E
M
A
L
A

HONDURAS

ST. LUCIA

BARB.

Guatemala ·
San Salvador ·

· Tegucigalpa

ST. VINCENT AND THE GRENADINES

GRENADA

EL SALVADOR

NICARAGUA

ARUBA (Neth.) NETH. ANT.

TRINIDAD AND TOBAGO

Managua ·

P A C I F I C   O C E A N

· San José

POLITICAL

Panamá ·

V E N E Z U E L A

COSTA RICA

G
U
Y
A
N
A

PANAMA

C O L O M B I A

A-530000-9J6   -3

©RMcN.

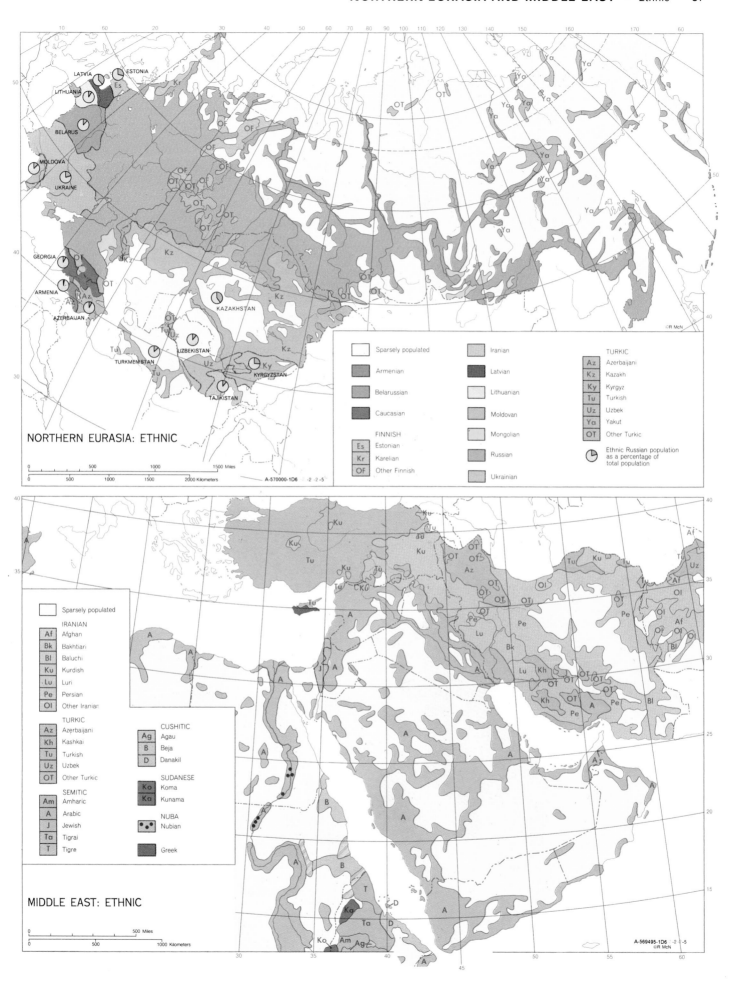

NORTHERN EURASIA: ETHNIC

Sparsely populated		Iranian		TURKIC	
Armenian		Latvian		Az	Azerbaijani
Belarussian		Lithuanian		Kz	Kazakh
Caucasian		Moldovan		Ky	Kyrgyz
FINNISH		Mongolian		Tu	Turkish
Es	Estonian	Russian		Uz	Uzbek
Kr	Karelian	Ukrainian		Ya	Yakut
OF	Other Finnish			OT	Other Turkic

Ethnic Russian population as a percentage of total population

A-570000-1D6   -2 -2 -5

MIDDLE EAST: ETHNIC

Sparsely populated

IRANIAN
Af Afghan
Bk Bakhtiari
Bl Baluchi
Ku Kurdish
Lu Luri
Pe Persian
OI Other Iranian

TURKIC
Az Azerbaijani
Kh Kashkai
Tu Turkish
Uz Uzbek
OT Other Turkic

SEMITIC
Am Amharic
A Arabic
J Jewish
Ta Tigrai
T Tigre

CUSHITIC
Ag Agau
B Beja
D Danakil

SUDANESE
Ko Koma
Ka Kunama

NUBA
Nubian

Greek

A-569495-1D6   -2 -1 -5
©R McN

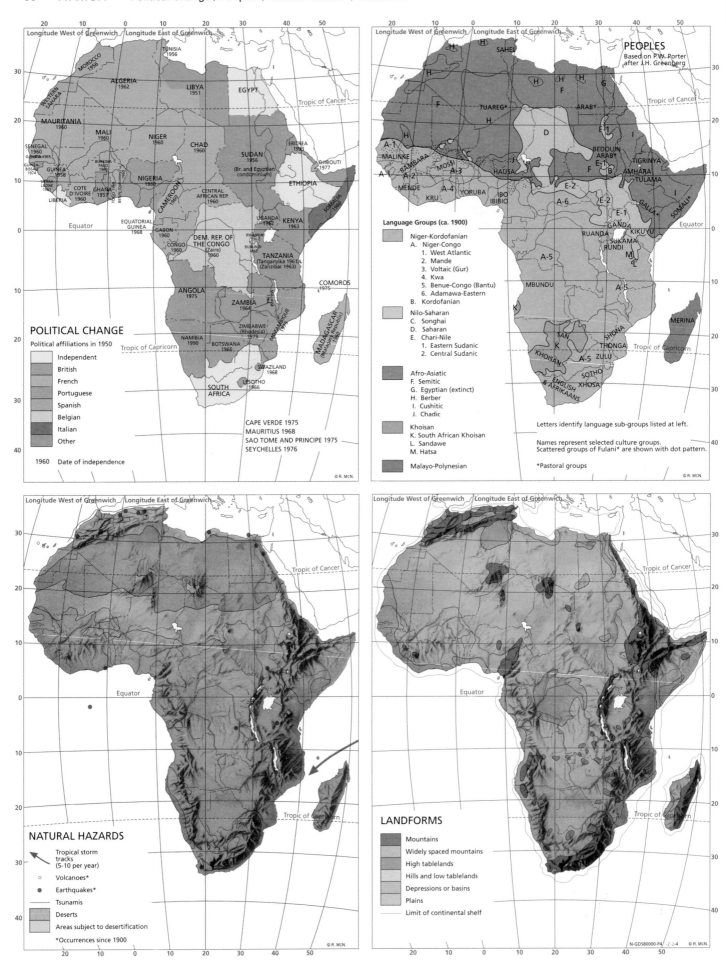

POLITICAL CHANGE

Political affiliations in 1950

Independent
British
French
Portuguese
Spanish
Belgian
Italian
Other

1960   Date of independence

CAPE VERDE 1975
MAURITIUS 1968
SAO TOME AND PRINCIPE 1975
SEYCHELLES 1976

PEOPLES

Based on P.W. Porter
after J.H. Greenberg

Language Groups (ca. 1900)

Niger-Kordofanian
  A. Niger-Congo
    1. West Atlantic
    2. Mande
    3. Voltaic (Gur)
    4. Kwa
    5. Benue-Congo (Bantu)
    6. Adamawa-Eastern
  B. Kordofanian

Nilo-Saharan
  C. Songhai
  D. Saharan
  E. Chari-Nile
    1. Eastern Sudanic
    2. Central Sudanic

Afro-Asiatic
  F. Semitic
  G. Egyptian (extinct)
  H. Berber
  I. Cushitic
  J. Chadic

Khoisan
  K. South African Khoisan
  L. Sandawe
  M. Hatsa

Malayo-Polynesian

Letters identify language sub-groups listed at left.

Names represent selected culture groups.
Scattered groups of Fulani* are shown with dot pattern.

*Pastoral groups

NATURAL HAZARDS

Tropical storm
tracks
(5–10 per year)

○ Volcanoes*
● Earthquakes*
      Tsunamis
      Deserts
      Areas subject to desertification

*Occurrences since 1900

LANDFORMS

Mountains
Widely spaced mountains
High tablelands
Hills and low tablelands
Depressions or basins
Plains
Limit of continental shelf

N-GDS80000-P4/ -2-2-4   © R. McN.

ORLD POLITICAL INFORMATION TABLE

table gives the area, population, population density, political status, al, and predominant languages for every country in the world. The ical units listed are categorized by political status in the form of rnment column of the table, as follows: A—independent countries; ternally independent political entities which are under the protection of her country in matters of defense and foreign affairs; C—colonies and r dependent political units; and D—the major administrative subdivisions of Australia, Canada, China, the United Kingdom, and the United States. For comparison, the table also includes the continents and the world. All footnotes appear at the end of the table.

The populations are estimates for January 1, 2004, made by Rand McNally on the basis of official data, United States Census Bureau estimates, and other available information. Area figures include inland water.

N OR POLITICAL DIVISION	Area Sq. Mi.	Est. Pop. 1/1/04	Pop. Per Sq. Mi.	Form of Government and Ruling Power	Capital	Predominant Languages	International Organizations
d Issas see Djibouti							
stan	251,773	29,205,000	116	Transitional A	Kābul	Dari, Pashto, Uzbek, Turkmen	UN
	11,700,000	866,305,000	74				
a	52,419	4,515,000	86	State (U.S.) D	Montgomery	English	
	663,267	650,000	1.0	State (U.S.) D	Juneau	English, indigenous	
	11,100	3,535,000	318	Republic A	Tiranë	Albanian, Greek	NATO/PP, UN
	255,541	3,215,000	13	Province (Canada) D	Edmonton	English	
	919,595	33,090,000	36	Republic A	Algiers (El Djazaïr)	Arabic, Berber dialects, French	AL, AU, OPEC, UN
n Samoa	77	58,000	753	Unincorporated territory (U.S.) C	Pago Pago	Samoan, English	
a	181	70,000	387	Parliamentary co-principality (Spanish and French) B	Andorra	Catalan, Spanish (Castilian), French, Portuguese	UN
	481,354	10,875,000	23	Republic A	Luanda	Portuguese, indigenous	AU, COMESA, UN
a	37	13,000	351	Overseas territory (U.K.) C	The Valley	English	
	53,668	61,215,000	1,141	Province (China) D	Hefei	Chinese (Mandarin)	
ca	5,400,000	([1])					
and Barbuda	171	68,000	398	Parliamentary state A	St. John's	English, local dialects	OAS, UN
(Macau)	6.9	445,000	64,493	Special administrative region (China) D	Macau (Aomen)	Chinese (Cantonese), Portuguese	
na	1,073,519	38,945,000	36	Republic A	Buenos Aires	Spanish, English, Italian, German, French	MERCOSUR, OAS, UN
	113,998	5,600,000	49	State (U.S.) D	Phoenix	English	
s	53,179	2,735,000	51	State (U.S.) D	Little Rock	English	
a	11,506	3,325,000	289	Republic A	Yerevan	Armenian, Russian	CIS, NATO/PP, UN
	75	71,000	947	Self-governing territory (Netherlands protection) B	Oranjestad	Dutch, Papiamento, English, Spanish	
on	34	1,000	29	Dependency (St. Helena) C	Georgetown	English	
	17,300,000	3,839,320,000	222				
a	2,969,910	19,825,000	6.7	Federal parliamentary state A	Canberra	English, indigenous	ANZUS, UN
an Capital Territory	911	325,000	357	Territory (Australia) D	Canberra	English	
	32,378	8,170,000	252	Federal republic A	Vienna (Wien)	German	EU, NATO/PP, UN
jan	33,437	7,850,000	235	Republic A	Baku (Bakı)	Azeri, Russian, Armenian	CIS, NATO/PP, UN
s	5,382	300,000	56	Parliamentary state A	Nassau	English, Creole	OAS, UN
	267	675,000	2,528	Monarchy A	Al Manāmah	Arabic, English, Persian, Urdu	AL, UN
desh	55,598	139,875,000	2,516	Republic A	Dkaha (Dacca)	Bangla, English	UN
s	166	280,000	1,687	Parliamentary state A	Bridgetown	English	OAS, UN
(Peking)	6,487	14,135,000	2,179	Autonomous city (China) D	Beijing (Peking)	Chinese (Mandarin)	
	80,155	10,315,000	129	Republic A	Minsk	Belarussian, Russian	CIS, NATO/PP, UN
ee Palau							
n	11,787	10,340,000	877	Constitutional monarchy A	Brussels (Bruxelles)	Dutch (Flemish), French, German	EU, NATO, UN
	8,867	270,000	30	Parliamentary state A	Belmopan	English, Spanish, Mayan, Garifuna, Creole	OAS, UN
	43,484	7,145,000	164	Republic A	Porto-Novo and Cotonou	French, Fon, Yoruba, indigenous	AU, UN
a	21	65,000	3,095	Overseas territory (U.K. protection) B	Hamilton	English, Portuguese	
	17,954	2,160,000	120	Monarchy (Indian protection) B	Thimphu	Dzongkha, Tibetan and Nepalese dialects	UN
	424,165	8,655,000	20	Republic A	La Paz and Sucre	Aymara, Quechua, Spanish	OAS, UN
and Herzegovina	19,767	4,000,000	202	Republic A	Sarajevo	Bosnian, Serbian, Croatian	UN
na	224,607	1,570,000	7.0	Republic A	Gaborone	English, Tswana	AU, UN
	3,300,172	183,080,000	55	Federal republic A	Brasília	Portuguese, Spanish, English, French	MERCOSUR, OAS, UN
Columbia	364,764	4,245,000	12	Province (Canada) D	Victoria	English	
ndian Ocean Territory	23	([1])		Overseas territory (U.K.) C		English	
Virgin Islands	58	22,000	379	Overseas territory (U.K.) C	Road Town	English	
	2,226	360,000	162	Monarchy A	Bandar Seri Begawan	Malay, English, Chinese	ASEAN, UN
a	42,855	7,550,000	176	Republic A	Sofia (Sofiya)	Bulgarian, Turkish	NATO, UN
a Faso	105,869	13,400,000	127	Republic A	Ouagadougou	French, indigenous	AU, UN
see Myanmar							
i	10,745	6,165,000	574	Republic A	Bujumbura	French, Kirundi, Swahili	AU, COMESA, UN
nia	163,696	35,590,000	217	State (U.S.) D	Sacramento	English	
dia	69,898	13,245,000	189	Constitutional monarchy A	Phnom Penh (Phnum Pénh)	Khmer, French, English	ASEAN, UN
oon	183,568	15,905,000	87	Republic A	Yaoundé	English, French, indigenous	AU, UN
	3,855,103	32,360,000	8.4	Federal parliamentary state A	Ottawa	English, French, other	NAFTA, NATO, OAS, UN
erde	1,557	415,000	267	Republic A	Praia	Portuguese, Crioulo	AU, UN
n Islands	102	43,000	422	Overseas territory (U.K.) C	George Town	English	
African Republic	240,536	3,715,000	15	Republic A	Bangui	French, Sango, indigenous	AU, UN
see Sri Lanka							
	495,755	9,395,000	19	Republic A	N'Djamena	Arabic, French, indigenous	AU, UN
el Islands	75	155,000	2,067	Two crown dependencies (U.K. protection)		English, French	
	291,930	15,745,000	54	Republic A	Santiago	Spanish	OAS, UN
excl. Taiwan)	3,690,045	1,298,720,000	352	Socialist republic A	Beijing (Peking)	Chinese dialects	UN
qing	31,815	31,600,000	993	Autonomous city (China) D	Chongqing (Chungking)	Chinese (Mandarin)	
as Island	52	400	7.7	External territory (Australia) C	Settlement	English, Chinese, Malay	
Keeling) Islands	5.4	600	111	External territory (Australia) C	West Island	English, Cocos-Malay	
bia	439,737	41,985,000	95	Republic A	Bogotá	Spanish	OAS, UN
do	104,094	4,565,000	44	State (U.S.) D	Denver	English	
os (excl. Mayotte)	863	640,000	742	Republic A	Moroni	Arabic, French, Shikomoro	AL, AU, COMESA, UN
	132,047	2,975,000	23	Republic A	Brazzaville	French, Lingala, Monokutuba, indigenous	AU, UN
Democratic Republic of the e)	905,446	57,445,000	63	Republic A	Kinshasa	French, Lingala, indigenous	AU, COMESA, UN
ticut	5,543	3,495,000	631	State (U.S.) D	Hartford	English	

REGION OR POLITICAL DIVISION	Area Sq. Mi.	Est. Pop. 1/1/04	Pop. Per Sq. Mi.	Form of Government and Ruling Power	Capital	Predominant Languages	International Organizations
Cook Islands	91	21,000	231	Self-governing territory (New Zealand protection) . . . . B	Avarua	English, Maori	
Costa Rica	19,730	3,925,000	199	Republic . . . . A	San José	Spanish, English	OAS, UN
Cote d'Ivoire (Ivory Coast)	124,504	17,145,000	138	Republic . . . . A	Abidjan and Yamoussoukro. .	French, Dioula and other indigenous	AU, UN
Croatia	21,829	4,430,000	203	Republic . . . . A	Zagreb	Croatian	NATO/PP, UN
Cuba	42,804	11,290,000	264	Socialist republic . . . . A	Havana (La Habana)	Spanish	OAS, UN
Cyprus	3,572	775,000	217	Republic . . . . A	Nicosia	Greek, Turkish, English	EU, UN
Czech Republic	30,450	10,250,000	337	Republic . . . . A	Prague (Praha)	Czech	EU, NATO, UN
Delaware	2,489	820,000	329	State (U.S.) . . . . D	Dover	English	
Denmark	16,640	5,405,000	325	Constitutional monarchy . . . . . A	Copenhagen (København) . . .	Danish	EU, NATO, UN
District of Columbia	68	565,000	8,309	Federal district (U.S.) . . . . D	Washington	English	
Djibouti	8,958	460,000	51	Republic . . . . A	Djibouti	French, Arabic, Somali, Afar	AL, AU, COMESA, U
Dominica	290	69,000	238	Republic . . . . A	Roseau	English, French	OAS, UN
Dominican Republic	18,730	8,775,000	468	Republic . . . . A	Santo Domingo	Spanish	OAS, UN
East Timor	5,743	1,010,000	176	Republic . . . . A	Dili	Portuguese, Tetum, Bahasa Indonesia (Malay), English	UN
Ecuador	109,484	13,840,000	126	Republic . . . . A	Quito	Spanish, Quechua, indigenous	OAS, UN
Egypt	386,662	75,420,000	195	Republic . . . . A	Cairo (Al Qāhirah)	Arabic	AL, AU, CAEU, COM UN
Ellice Islands see Tuvalu							
El Salvador	8,124	6,530,000	804	Republic . . . . A	San Salvador	Spanish, Nahua	OAS, UN
England	50,356	50,360,000	1,000	Administrative division (U.K.) . . . D	London	English	
Equatorial Guinea	10,831	515,000	48	Republic . . . . A	Malabo	French, Spanish, indigenous, English	AU, UN
Eritrea	45,406	4,390,000	97	Republic . . . . A	Asmera	Afar, Arabic, Tigre, Kunama, Tigrinya, other	AU, COMESA, UN
Estonia	17,462	1,405,000	80	Republic . . . . A	Tallinn	Estonian, Russian, Ukrainian, Finnish, other	EU, NATO, UN
Ethiopia	426,373	67,210,000	158	Federal republic . . . . A	Addis Ababa (Adis Abeba) . . .	Amharic, Tigrinya, Orominga, Guaraginga, Somali, Arabic	AU, COMESA, UN
Europe	3,800,000	729,330,000	192				
Falkland Islands (²)	4,700	3,000	0.6	Overseas territory (U.K.) C	Stanley	English	
Faroe Islands	540	47,000	87	Self-governing territory (Danish protection). . B	Tórshavn	Danish, Faroese	
Fiji	7,056	875,000	124	Republic . . . . A	Suva	English, Fijian, Hindustani	UN
Finland	130,559	5,210,000	40	Republic . . . . A	Helsinki (Helsingfors)	Finnish, Swedish, Sami, Russian	EU, NATO/PP, UN
Florida	65,755	17,070,000	260	State (U.S.) . . . . D	Tallahassee	English	
France (excl. Overseas Departments)	208,482	60,305,000	289	Republic . . . . A	Paris	French	EU, NATO, UN
French Guiana	32,253	190,000	5.9	Overseas department (France) . . C	Cayenne	French	
French Polynesia	1,544	265,000	172	Overseas territory (France) C	Papeete	French, Tahitian	
Fujian	46,332	35,495,000	766	Province (China) . . . . D	Fuzhou	Chinese dialects	
Gabon	103,347	1,340,000	13	Republic . . . . A	Libreville	French, Fang, indigenous	AU, UN
Gambia, The	4,127	1,525,000	370	Republic . . . . A	Banjul	English, Malinke, Wolof, Fula, indigenous	AU, UN
Gansu	173,746	26,200,000	151	Province (China) . . . . D	Lanzhou	Chinese (Mandarin), Mongolian, Tibetan dialects	
Gaza Strip	139	1,300,000	9,353	Israeli territory with limited self-government. . . .		Arabic, Hebrew	(⁷)
Georgia	59,425	8,710,000	147	State (U.S.) . . . . D	Atlanta	English	
Georgia	26,911	4,920,000	183	Republic . . . . A	Tbilisi	Georgian, Russian, Armenian, Azeri, other	NATO/PP, UN
Germany	137,847	82,415,000	598	Federal republic . . . . A	Berlin	German	EU, NATO, UN
Ghana	92,098	20,615,000	224	Republic . . . . A	Accra	English, Akan and other indigenous	AU, UN
Gibraltar (⁷)	2.3	28,000	12,174	Overseas territory (U.K.) C	Gibraltar	English, Spanish, Italian, Portuguese	
Gilbert Islands see Kiribati							
Golan Heights	454	37,000	81	Occupied by Israel		Arabic, Hebrew	
Great Britain see United Kingdom							
Greece	50,949	10,635,000	209	Republic . . . . A	Athens (Athina)	Greek, English, French	EU, NATO, UN
Greenland	836,331	56,000	0.07	Self-governing territory (Danish protection). . B	Godthåb (Nuuk)	Danish, Greenlandic, English	
Grenada	133	89,000	669	Parliamentary state. . . . . A	St. George's	English, French	OAS, UN
Guadeloupe (incl. Dependencies)	687	440,000	640	Overseas department (France) . . C	Basse-Terre	French, Creole	
Guam	212	165,000	778	Unincorporated territory (U.S.). . . C	Hagåtña (Agana)	English, Chamorro, Japanese	
Guangdong	68,649	88,375,000	1,287	Province (China) . . . . D	Guangzhou (Canton)	Chinese dialects, Miao-Yao	
Guangxi Zhuangzu	91,236	45,905,000	503	Autonomous region (China) . . . . D	Nanning	Chinese dialects, Thai, Miao-Yao	
Guatemala	42,042	14,095,000	335	Republic . . . . A	Guatemala	Spanish, indigenous	OAS, UN
Guernsey (incl. Dependencies)	30	65,000	2,167	Crown dependency (U.K. protection) . . B	St. Peter Port	English, French	
Guinea	94,926	9,135,000	96	Republic . . . . A	Conakry	French, indigenous	AU, UN
Guinea-Bissau	13,948	1,375,000	99	Republic . . . . A	Bissau	Portuguese, Crioulo, indigenous	AU, UN
Guizhou	65,637	36,045,000	549	Province (China) . . . . D	Guiyang	Chinese (Mandarin), Thai, Miao-Yao	
Guyana	83,000	705,000	8.5	Republic . . . . A	Georgetown	English, indigenous, Creole, Hindi, Urdu	OAS, UN
Hainan	13,205	8,050,000	610	Province (China) . . . . D	Haikou	Chinese, Min, Tai	
Haiti	10,714	7,590,000	708	Republic . . . . A	Port-au-Prince	Creole, French	OAS, UN
Hawaii	10,931	1,260,000	115	State (U.S.) . . . . D	Honolulu	English, Hawaiian, Japanese	
Hebei	73,359	68,965,000	940	Province (China) . . . . D	Shijiazhuang	Chinese (Mandarin)	
Heilongjiang	181,082	37,725,000	208	Province (China) . . . . D	Harbin	Chinese dialects, Mongolian, Tungus. . .	
Henan	64,479	94,655,000	1,468	Province (China) . . . . D	Zhengzhou	Chinese (Mandarin)	
Holland see Netherlands							
Honduras	43,277	6,745,000	156	Republic . . . . A	Tegucigalpa	Spanish, indigenous	OAS, UN
Hubei	72,356	61,645,000	852	Province (China) . . . . D	Wuhan	Chinese dialects	
Hunan	81,082	65,855,000	812	Province (China) . . . . D	Changsha	Chinese dialects, Miao-Yao	
Hungary	35,919	10,045,000	280	Republic . . . . A	Budapest	Hungarian	EU, NATO, UN
Iceland	39,769	280,000	7.0	Republic . . . . A	Reykjavik	Icelandic, English, other	EFTA, NATO, UN
Idaho	83,570	1,370,000	16	State (U.S.) . . . . D	Boise	English	
Illinois	57,914	12,690,000	219	State (U.S.) . . . . D	Springfield	English	
India (incl. part of Jammu and Kashmir)	1,222,510	1,057,415,000	865	Federal republic . . . . A	New Delhi	English, Hindi, Telugu, Bengali, indigenous	UN
Indiana	36,418	6,215,000	171	State (U.S.) . . . . D	Indianapolis	English	
Indonesia	735,310	236,680,000	322	Republic . . . . A	Jakarta	Bahasa Indonesia (Malay), English, Dutch, indigenous	ASEAN, OPEC, UN
Iowa	56,272	2,955,000	53	State (U.S.) . . . . D	Des Moines	English	
Iran	636,372	68,650,000	108	Islamic republic . . . . A	Tehrān	Persian, Turkish dialects, Kurdish, other	OPEC, UN
Iraq	169,235	25,025,000	148	Republic . . . . A	Baghdād	Arabic, Kurdish, Assyrian, Armenian	AL, CAEU, OPEC, UN
Ireland	27,133	3,945,000	145	Republic . . . . A	Dublin (Baile Átha Cliath) . . .	English, Irish Gaelic	EU, NATO/PP, UN
Isle of Man	221	74,000	335	Crown dependency (U.K. protection) . . . . . . B	Douglas	English, Manx Gaelic	

N OR POLITICAL DIVISION	Area Sq. Mi.	Est. Pop. 1/1/04	Pop. Per Sq. Mi.	Form of Government and Ruling Power	Capital	Predominant Languages	International Organizations
xcl. Occupied Areas)	8,019	6,160,000	768	Republic A	Jerusalem (Yerushalayim)	Hebrew, Arabic	UN
	116,342	58,030,000	499	Republic A	Rome (Roma)	Italian, German, French, Slovene	EU, NATO, UN
ast see Cote d'Ivoire							
	4,244	2,705,000	637	Parliamentary state A	Kingston	English, Creole	OAS, UN
	145,850	127,285,000	873	Constitutional monarchy A	Tōkyō	Japanese	UN
	45	90,000	2,000	Crown dependency (U.K. protection) B	St. Helier	English, French	
	39,614	76,065,000	1,920	Province (China) D	Nanjing (Nanking)	Chinese dialects	
	64,325	42,335,000	658	Province (China) D	Nanchang	Chinese dialects	
	72,201	27,895,000	386	Province (China) D	Changchun	Chinese (Mandarin), Mongolian, Korean	
	34,495	5,535,000	160	Constitutional monarchy A	'Ammān	Arabic	AL, CAEU, UN
	82,277	2,730,000	33	State (U.S.) D	Topeka	English	
stan	1,049,156	16,780,000	16	Republic A	Astana (Aqmola)	Kazakh, Russian	CIS, NATO/PP, UN
ky	40,409	4,130,000	102	State (U.S.) D	Frankfort	English	
	224,961	31,840,000	142	Republic A	Nairobi	English, Swahili, indigenous	AU, COMESA, UN
	313	100,000	319	Republic A	Bairiki	English, I-Kiribati	UN
North	46,540	22,585,000	485	Socialist republic A	P'yŏngyang	Korean	UN
South	38,328	48,450,000	1,264	Republic A	Seoul (Sŏul)	Korean	UN
	6,880	2,220,000	323	Constitutional monarchy A	Kuwait (Al Kuwayt)	Arabic, English	AL, CAEU, OPEC, UN
tan	77,182	4,930,000	64	Republic A	Bishkek	Kirghiz, Russian	CIS, NATO/PP, UN
	91,429	5,995,000	66	Socialist republic A	Viangchan (Vientiane)	Lao, French, English	ASEAN, UN
	24,942	2,340,000	94	Republic A	Rīga	Latvian, Lithuanian, Russian, other	EU, NATO, UN
n	4,016	3,755,000	935	Republic A	Beirut (Bayrūt)	Arabic, French, Armenian, English	AL, UN
	11,720	1,865,000	159	Constitutional monarchy A	Maseru	English, Sesotho, Zulu, Xhosa	AU, UN
g	56,255	43,340,000	770	Province (China) D	Shenyang (Mukden)	Chinese (Mandarin), Mongolian	
	43,000	3,345,000	78	Republic A	Monrovia	English, indigenous	AU, UN
	679,362	5,565,000	8.2	Socialist republic A	Tripoli (Ṭarābulus)	Arabic	AL, AU, CAEU, OPEC, UN
nstein	62	33,000	532	Constitutional monarchy A	Vaduz	German	EFTA, UN
ia	25,213	3,590,000	142	Republic A	Vilnius	Lithuanian, Polish, Russian	EU, NATO, UN
na	51,840	4,510,000	87	State (U.S.) D	Baton Rouge	English	
bourg	999	460,000	460	Constitutional monarchy A	Luxembourg	French, Luxembourgish, German	EU, NATO, UN
onia	9,928	2,065,000	208	Republic A	Skopje	Macedonian, Albanian, other	NATO/PP, UN
ascar	226,658	17,235,000	76	Republic A	Antananarivo	French, Malagasy	AU, COMESA, UN
	35,385	1,310,000	37	State (U.S.) D	Augusta	English	
	45,747	11,780,000	258	Republic A	Lilongwe	Chichewa, English, indigenous	AU, COMESA, UN
a	127,320	23,310,000	183	Federal constitutional monarchy A	Kuala Lumpur and Putrajaya ([5])	Bahasa Melayu, Chinese dialects, English, other	ASEAN, UN
es	115	335,000	2,913	Republic A	Male'	Dhivehi	UN
	478,841	11,790,000	25	Republic A	Bamako	French, Bambara, indigenous	AU, UN
	122	400,000	3,279	Republic A	Valletta	English, Maltese	EU, UN
ba	250,116	1,190,000	4.8	Province (Canada) D	Winnipeg	English	
ll Islands	70	57,000	814	Republic (U.S. protection) B	Majuro (island)	English, indigenous, Japanese	UN
que	425	430,000	1,012	Overseas department (France) C	Fort-de-France	French, Creole	
nd	12,407	5,525,000	445	State (U.S.) D	Annapolis	English	
husetts	10,555	6,455,000	612	State (U.S.) D	Boston	English	
ania	397,956	2,955,000	7.4	Republic A	Nouakchott	Arabic, Wolof, Pular, Soninke, French	AL, AU, CAEU, UN
us (incl. Dependencies)	788	1,215,000	1,542	Republic A	Port Louis	English, French, Creole, other	AU, COMESA, UN
e ([6])	144	180,000	1,250	Departmental collectivity (France) C	Mamoutzou	French, Swahili (Mahorian)	
	758,452	104,340,000	138	Federal republic A	Mexico City (Ciudad de México)	Spanish, indigenous	NAFTA, OAS, UN
an	96,716	10,110,000	105	State (U.S.) D	Lansing	English	
esia, Federated States of	271	110,000	406	Republic (U.S. protection) B	Palikir	English, indigenous	UN
y Islands	2.0	([1])		Unincorporated territory (U.S.) C		English	
ota	86,939	5,075,000	58	State (U.S.) D	St. Paul	English	
ppi	48,430	2,890,000	60	State (U.S.) D	Jackson	English	
ri	69,704	5,720,000	82	State (U.S.) D	Jefferson City	English	
va	13,070	4,440,000	340	Republic A	Chişinău (Kishinev)	Romanian (Moldovan), Russian, Gagauz	CIS, NATO/PP, UN
o	0.8	32,000	40,000	Constitutional monarchy A	Monaco	French, English, Italian, Monegasque	UN
lia	604,829	2,730,000	4.5	Republic A	Ulan Bator (Ulaanbaatar)	Khalkha Mongol, Turkish dialects, Russian	UN
na	4,095	920,000	225	State (U.S.) D	Helena	English	
errat	39	9,000	231	Overseas territory (U.K.) C	Plymouth	English	
co (excl. Western Sahara)	172,414	31,950,000	185	Constitutional monarchy A	Rabat	Arabic, Berber dialects, French	AL, UN
bique	309,496	18,695,000	60	Republic A	Maputo	Portuguese, indigenous	AU, UN
nar (Burma)	261,228	42,620,000	163	Provisional military government A	Rangoon (Yangon)	Burmese, indigenous	ASEAN, UN
a	317,818	1,940,000	6.1	Republic A	Windhoek	English, Afrikaans, German, indigenous	AU, COMESA, UN
	8.1	13,000	1,605	Republic A	Yaren District	Nauruan, English	UN
ka	77,354	1,745,000	23	State (U.S.) D	Lincoln	English	
ngol (Inner Mongolia)	456,759	24,295,000	53	Autonomous region (China) D	Hohhot	Mongolian	
	56,827	26,770,000	471	Constitutional monarchy A	Kathmandu	Nepali, indigenous	UN
lands	16,164	16,270,000	1,007	Constitutional monarchy A	Amsterdam and The Hague ('s-Gravenhage)	Dutch, Frisian	EU, NATO, UN
lands Antilles	309	215,000	696	Self-governing territory (Netherlands protection) B	Willemstad	Dutch, Papiamento, English, Spanish	
a	110,561	2,250,000	20	State (U.S.) D	Carson City	English	
runswick	28,150	770,000	27	Province (Canada) D	Fredericton	English, French	
aledonia	7,172	210,000	29	Territorial collectivity (France) C	Nouméa	French, indigenous	
undland and Labrador	156,453	535,000	3.4	Province (Canada) D	St. John's	English	
ampshire	9,350	1,290,000	138	State (U.S.) D	Concord	English	
ebrides see Vanuatu							
ersey	8,721	8,665,000	994	State (U.S.) D	Trenton	English	
lexico	121,590	1,880,000	15	State (U.S.) D	Santa Fe	English, Spanish	
outh Wales	309,129	6,665,000	22	State (Australia) D	Sydney	English	
ork	54,556	19,245,000	353	State (U.S.) D	Albany	English	
ealand	104,454	3,975,000	38	Parliamentary state A	Wellington	English, Maori	ANZUS, UN
gua	50,054	5,180,000	103	Republic A	Managua	Spanish, English, indigenous	OAS, UN
	489,192	11,210,000	23	Repubic A	Niamey	French, Hausa, Djerma, indigenous	AU, UN
a	356,669	135,570,000	380	Transitional military government A	Abuja	English, Hausa, Fulani, Yorbua, Ibo, indigenous	AU, OPEC, UN
a Huizu	25,637	5,745,000	224	Autonomous region (China) D	Yinchuan	Chinese (Mandarin)	
	100	2,000	20	Self-governing territory (New Zealand protection) B	Alofi	Niuean, English	
k Island	14	2,000	143	External territory (Australia) C	Kingston	English, Norfolk	

REGION OR POLITICAL DIVISION	Area Sq. Mi.	Est. Pop. 1/1/04	Pop. Per Sq. Mi.	Form of Government and Ruling Power	Capital	Predominant Languages	International Organizations
North America	9,500,000	505,780,000	53				
North Carolina	53,819	8,430,000	157	State (U.S.) D	Raleigh	English	
North Dakota	70,700	635,000	9.0	State (U.S.) D	Bismarck	English	
Northern Ireland	5,242	1,725,000	329	Administrative division (U.K.) D	Belfast	English	
Northern Mariana Islands	179	77,000	430	Commonwealth (U.S. protection) B	Saipan (island)	English, Chamorro, Carolinian	
Northern Territory	520,902	200,000	0.4	Territory (Australia) D	Darwin	English, indigenous	
Northwest Territories	519,735	43,000	0.08	Territory (Canada) D	Yellowknife	English, indigenous	
Norway (incl. Svalbard and Jan Mayen)	125,050	4,565,000	37	Constitutional monarchy A	Oslo	Norwegian, Sami, Finnish	EFTA, NATO, UN
Nova Scotia	21,345	965,000	45	Province (Canada) D	Halifax	English	
Nunavut	808,185	30,000	0.04	Territory (Canada) D	Iqaluit	English, indigenous	
Oceania (incl. Australia)	3,300,000	32,170,000	9.7				
Ohio	44,825	11,470,000	256	State (U.S.) D	Columbus	English	
Oklahoma	69,898	3,520,000	50	State (U.S.) D	Oklahoma City	English	
Oman	119,499	2,855,000	24	Monarchy A	Muscat (Masqat)	Arabic, English, Baluchi, Urdu, Indian dialects	AL, UN
Ontario	415,599	12,495,000	30	Province (Canada) D	Toronto	English	
Oregon	98,381	3,570,000	36	State (U.S.) D	Salem	English	
Pakistan (incl. part of Jammu and Kashmir)	339,732	152,210,000	448	Federal Islamic republic A	Islāmābād	English, Urdu, Punjabi, Sindhi, Pashto, other	UN
Palau (Belau)	188	20,000	106	Republic (U.S. protection) B	Koror and Melekeok (¹)	Angaur, English, Japanese, Palauan, Sonsorolese, Tobi	UN
Panama	29,157	2,980,000	102	Republic A	Panamá	Spanish, English	OAS, UN
Papua New Guinea	178,704	5,360,000	30	Parliamentary state A	Port Moresby	English, Motu, Pidgin, indigenous	UN
Paraguay	157,048	6,115,000	39	Republic A	Asunción	Guarani, Spanish	MERCOSUR, OAS, U
Pennsylvania	46,055	12,400,000	269	State (U.S.) D	Harrisburg	English	
Peru	496,225	28,640,000	58	Republic A	Lima	Quechua, Spanish, Aymara	OAS, UN
Philippines	115,831	85,430,000	738	Republic A	Manila	English, Filipino, indigenous	ASEAN, UN
Pitcairn Islands (incl. Dependencies)	19	100	5.3	Overseas territory (U.K.) C	Adamstown	English, Pitcairnese	
Poland	120,728	38,625,000	320	Republic A	Warsaw (Warszawa)	Polish	EU, NATO, UN
Portugal	35,516	10,110,000	285	Republic A	Lisbon (Lisboa)	Portuguese, Mirandese	EU, NATO, UN
Prince Edward Island	2,185	140,000	64	Province (Canada) D	Charlottetown	English	
Puerto Rico	3,515	3,890,000	1,107	Commonwealth (U.S. protection) B	San Juan	Spanish, English	
Qatar	4,412	830,000	188	Monarchy A	Ad Dawḩah (Doha)	Arabic	AL, OPEC, UN
Qinghai	277,994	5,295,000	19	Province (China) D	Xining	Tibetan dialects, Mongolian, Turkish dialects, Chinese (Mandarin)	
Quebec	595,391	7,675,000	13	Province (Canada) D	Québec	French, English	
Queensland	668,208	3,785,000	5.7	State (Australia) D	Brisbane	English	
Reunion	969	760,000	784	Overseas department (France) C	Saint-Denis	French, Creole	
Rhode Island	1,545	1,080,000	699	State (U.S.) D	Providence	English	
Rhodesia see Zimbabwe							
Romania	91,699	22,370,000	244	Republic A	Bucharest (Bucureşti)	Romanian, Hungarian, German	NATO, UN
Russia	6,592,849	144,310,000	22	Federal republic A	Moscow (Moskva)	Russian, other	CIS, NATO/PP, UN
Rwanda	10,169	7,880,000	775	Republic A	Kigali	English, French, Kinyarwanda, Kiswahili	AU, COMESA, UN
St. Helena (incl. Dependencies)	121	7,500	62	Overseas territory (U.K.) C	Jamestown	English	
St. Kitts and Nevis	101	39,000	386	Parliamentary state A	Basseterre	English	OAS, UN
St. Lucia	238	165,000	693	Parliamentary state A	Castries	English, French	OAS, UN
St. Pierre and Miquelon	93	7,000	75	Territorial collectivity (France) C	Saint-Pierre	French	
St. Vincent and the Grenadines	150	115,000	767	Parliamentary state A	Kingstown	English, French	OAS, UN
Samoa	1,093	180,000	165	Constitutional monarchy A	Apia	English, Samoan	UN
San Marino	24	28,000	1,167	Republic A	San Marino	Italian	UN
Sao Tome and Principe	372	180,000	484	Republic A	São Tomé	Portuguese	AU, UN
Saskatchewan	251,366	1,025,000	4.1	Province (Canada) D	Regina	English	
Saudi Arabia	830,000	24,690,000	30	Monarchy A	Riyadh (Ar Riyāḑ)	Arabic	AL, OPEC, UN
Scotland	30,167	5,135,000	170	Administrative division (U.K.) D	Edinburgh	English, Scots Gaelic	
Senegal	75,951	10,715,000	141	Republic A	Dakar	French, Wolof and other indigenous	AU, UN
Serbia and Montenegro (Yugoslavia)	39,449	10,660,000	270	Republic A	Belgrade (Beograd)	Serbian, Albanian	UN
Seychelles	176	81,000	460	Republic A	Victoria	English, French, Creole	AU, COMESA, UN
Shaanxi	79,151	36,865,000	466	Province (China) D	Xi'an (Sian)	Chinese (Mandarin)	
Shandong	59,074	92,845,000	1,572	Province (China) D	Jinan	Chinese (Mandarin)	
Shanghai	2,394	17,120,000	7,151	Autonomous city (China) D	Shanghai	Chinese (Wu)	
Shanxi	60,232	33,715,000	560	Province (China) D	Taiyuan	Chinese (Mandarin)	
Sichuan	188,263	85,175,000	452	Province (China) D	Chengdu	Chinese (Mandarin), Tibetan dialects, Miao-Yao	
Sierra Leone	27,699	5,815,000	210	Republic A	Freetown	English, Krio, Mende, Temne, indigenous	AU, UN
Singapore	264	4,685,000	17,746	Republic A	Singapore	Chinese (Mandarin), English, Malay, Tamil	ASEAN, UN
Slovakia	18,924	5,420,000	286	Republic A	Bratislava	Slovak, Hungarian	EU, NATO, UN
Slovenia	7,821	1,935,000	247	Republic A	Ljubljana	Slovenian, Croatian, Serbian	EU, NATO, UN
Solomon Islands	10,954	515,000	47	Parliamentary state A	Honiara	English, indigenous	UN
Somalia	246,201	8,165,000	33	Transitional A	Mogadishu (Muqdisho)	Arabic, Somali, English, Italian	AL, AU, CAEU, UN
South Africa	470,693	42,770,000	91	Republic A	Pretoria, Cape Town, and Bloemfontein	Afrikaans, English, Xhosa, Zulu, other indigenous	AU, UN
South America	6,900,000	366,600,000	53				
South Australia	379,724	1,525,000	4.0	State (Australia) D	Adelaide	English	
South Carolina	32,020	4,160,000	130	State (U.S.) D	Columbia	English	
South Dakota	77,117	765,000	9.9	State (U.S.) D	Pierre	English	
South Georgia and the South Sandwich Islands (²)	1,450	(¹)	.....	Overseas territory (U.K.) C		English	
South West Africa see Namibia							
Spain	194,885	40,250,000	207	Constitutional monarchy A	Madrid	Spanish (Castilian), Catalan, Galician, Basque	EU, NATO, UN
Spanish North Africa (³)	12	140,000	11,667	Five possessions (Spain) C		Spanish, Arabic, Berber dialects	
Spanish Sahara see Western Sahara							
Sri Lanka	25,332	19,825,000	783	Socialist republic A	Colombo and Sri Jayewardenepura Kotte	English, Sinhala, Tamil	UN
Sudan	967,500	38,630,000	40	Provisional military government A	Khartoum (Al Kharţūm)	Arabic, Nubian, and other indigenous, English	AL, AU, CAEU, COM UN
Suriname	63,037	435,000	6.9	Republic A	Paramaribo	Dutch, Sranan Tongo, English, Hindustani, Javanese	OAS, UN

...N OR POLITICAL DIVISION	Area Sq. Mi.	Est. Pop. 1/1/04	Pop. Per Sq. Mi.	Form of Government and Ruling Power	Capital	Predominant Languages	International Organizations
...nd	6,704	1,165,000	174	Monarchy ... A	Mbabane and Lobamba	English, siSwati	AU, COMESA, UN
	173,732	8,980,000	52	Constitutional monarchy ... A	Stockholm	Swedish, Sami, Finnish	EU, NATO/PP, UN
...land	15,943	7,430,000	466	Federal republic ... A	Bern (Berne)	German, French, Italian, Romansch	EFTA, NATO/PP, UN
	71,498	17,800,000	249	Republic ... A	Damascus (Dimashq)	Arabic, Kurdish, Armenian, Aramaic, Circassian	AL, CAEU, UN
	13,901	22,675,000	1,631	Republic ... A	T'aipei	Chinese (Mandarin), Taiwanese (Min), Hakka	
...an	55,251	6,935,000	126	Republic ... A	Dushanbe	Tajik, Russian	CIS, NATO/PP, UN
...ia	364,900	36,230,000	99	Republic ... A	Dar es Salaam and Dodoma	English, Swahili, indigenous	AU, UN
...ia	26,409	475,000	18	State (Australia) ... D	Hobart	English	
...see	42,143	5,860,000	139	State (U.S.) ... D	Nashville	English	
	268,581	22,185,000	83	State (U.S.) ... D	Austin	English, Spanish	
...d	198,115	64,570,000	326	Constitutional monarchy ... A	Bangkok (Krung Thep)	Thai, indigenous	ASEAN, UN
...(Tientsin)	4,363	10,235,000	2,346	Autonomous city (China) ... D	Tianjin (Tientsin)	Chinese (Mandarin)	
	21,925	5,495,000	251	Republic ... A	Lomé	French, Ewe, Mina, Kabye, Dagomba	AU, UN
...u	4.6	1,500	326	Island territory (New Zealand) ... C		English, Tokelauan	
	251	110,000	438	Constitutional monarchy ... A	Nuku'alofa	Tongan, English	UN
...d and Tobago	1,980	1,100,000	556	Republic ... A	Port of Spain	English, Hindi, French, Spanish, Chinese	OAS, UN
...da Cunha	40	300	7.5	Dependency (St. Helena) ... C	Edinburgh	English	
	63,170	9,980,000	158	Republic ... A	Tunis	Arabic, French	AL, AU, UN
	302,541	68,505,000	226	Republic ... A	Ankara	Turkish, Kurdish, Arabic, Armenian, Greek	NATO, UN
...nistan	188,457	4,820,000	26	Republic ... A	Ashgabat (Ashkhabad)	Turkmen, Russian, Uzbek	CIS, NATO/PP, UN
...nd Caicos Islands	166	20,000	120	Overseas territory (U.K.) ... C	Grand Turk	English	
	10	11,000	1,100	Parliamentary state ... A	Funafuti	Tuvaluan, English, Samoan, I-Kiribati	UN
...a	93,065	26,010,000	279	Republic ... A	Kampala	English, Luganda, Swahili, indigenous, Arabic	AU, COMESA, UN
...e	233,090	47,890,000	205	Republic ... A	Kiev (Kyïv)	Ukrainian, Russian, Romanian, Polish, Hungarian	CIS, NATO/PP, UN
...Arab Emirates	32,278	2,505,000	78	Federation of monarchs ... A	Abū Ẓaby (Abu Dhabi)	Arabic, Persian, English, Hindi, Urdu	AL, CAEU, OPEC, UN
...Kingdom	93,788	60,185,000	642	Constitutional monarchy ... A	London	English, Welsh, Scots Gaelic	EU, NATO, UN
...States	3,794,083	291,680,000	77	Federal republic ... A	Washington	English, Spanish	ANZUS, NAFTA, NATO, OAS, UN
...Volta see Burkina Faso							
...ay	67,574	3,425,000	51	Republic ... A	Montevideo	Spanish	MERCOSUR, OAS, UN
	84,899	2,360,000	28	State (U.S.) ... D	Salt Lake City	English	
...stan	172,742	26,195,000	152	Republic ... A	Tashkent (Toshkent)	Uzbek, Russian, Tajik	CIS, NATO/PP, UN
...tu	4,707	200,000	42	Republic ... A	Port Vila	Bislama, English, French	UN
...City	0.2	900	4,500	Ecclesiastical state ... A	Vatican City	Italian, Latin, French, other	
...uela	352,145	24,835,000	71	Federal republic ... A	Caracas	Spanish, indigenous	OAS, OPEC, UN
...nt	9,614	620,000	64	State (U.S.) ... D	Montpelier	English	
...a	87,807	4,905,000	56	State (Australia) ... D	Melbourne	English	
...m	128,066	82,150,000	641	Socialist republic ... A	Hanoi	Vietnamese, English, French, Chinese, Khmer, indigenous	ASEAN, UN
...a	42,774	7,410,000	173	State (U.S.) ... D	Richmond	English	
...Islands (U.S.)	134	110,000	821	Unincorporated territory (U.S.) ... C	Charlotte Amalie	English, Spanish, Creole	
...Island	3.0	110,000	(')	Unincorporated territory (U.S.) ... C		English	
	8,023	2,965,000	370	Administrative division (U.K.) ... D	Cardiff	English, Welsh Gaelic	
...and Futuna	99	16,000	162	Overseas territory (France) ... C	Mata-Utu	French, Wallisian	
...ngton	71,300	6,150,000	86	State (U.S.) ... D	Olympia	English	
...ank (incl. Jericho and Jerusalem)	2,263	2,275,000	1,005	Israeli territory with limited self-government ...		Arabic, Hebrew	(')
...n Australia	976,792	1,945,000	2.0	State (Australia) ... D	Perth	English	
...n Sahara	102,703	265,000	2.6	Occupied by Morocco ... C		Arabic	
...Virginia	24,230	1,815,000	75	State (U.S.) ... D	Charleston	English	
...nsin	65,498	5,490,000	84	State (U.S.) ... D	Madison	English	
...ing	97,814	505,000	5.2	State (U.S.) ... D	Cheyenne	English	
...gang (Hong Kong)	425	7,440,000	17,506	Special administrative region (China) ... D	Hong Kong (Xianggang)	Chinese (Cantonese), English	
...g Uygur (Sinkiang)	617,764	19,685,000	32	Autonomous region (China) ... D	Ürümqi	Turkish dialects, Mongolian, Tungus, English	
...g (Tibet)	471,045	2,680,000	5.7	Autonomous region (China) ... D	Lhasa	Tibetan dialects	
...n	203,850	19,680,000	97	Republic ... A	Şan'ā' (Sanaa)	Arabic	AL, CAEU, UN
...avia see Serbia and Montenegro							
...Territory	186,272	32,000	0.2	Territory (Canada) ... D	Whitehorse	English, Inuktitut, indigenous	
...n	152,124	43,850,000	288	Province (China) ... D	Kunming	Chinese (Mandarin), Tibetan dialects, Khmer, Miao-Yao	
...ee Congo, Democratic ...ublic of the							
...a	290,586	10,385,000	36	Republic ... A	Lusaka	English, indigenous	AU, COMESA, UN
...ing	39,305	47,830,000	1,217	Province (China) ... D	Hangzhou	Chinese dialects	
...bwe	150,873	12,630,000	84	Republic ... A	Harare (Salisbury)	English, indigenous	AU, COMESA, UN
...D	57,900,000	6,339,505,000	109				

...e, or not applicable
...permanent population
...imed by Argentina
...imed by Spain
...e Palestinian Liberation Organization (PLO) is a member of AL and CAEU
...ure capital
...imed by Comoros
...mprises Ceuta, Melilla, and several small islands

	Arab League (League of Arab States)
...5	Australia-New Zealand-U.S. Security Treaty
	Association of Southeast Asian Nations
	African Union
	Council of Arab Unity
	Commonwealth of Independent States
...SA	Common Market for Eastern and Southern Africa
	European Free Trade Association
	European Union
...OSUR	Southern Common Market
...A	North American Free Trade Agreement
	North Atlantic Treaty Organization
...PP	NATO-Partnership for Peace Program
	Organization of American States
	Organization of Petroleum Exporting Countries

WORLD DEMOGRAPHIC TABLE

CONTINENT/Country	Population Estimate 2004	Pop. Per Sq. Mile 2004	Percent Urban[1] 2001	Crude Birth Rate per 1,000[2] 2003	Crude Death Rate per 1,000[2] 2003	Natural Increase Percent[2] 2003	Fertility Rate (Children born/Woman)[3] 2003	Infant Mortality Rate per 1,000[3] 2003	Median Age[2] 2002	Life Expectancy Male[2] 2003	Ex
NORTH AMERICA											
Bahamas	300,000	56	64.7	19	9	1.0%	2	26	27	62	
Belize	270,000	30	48.1	30	6	2.4%	4	27	19	65	
Canada	32,360,000	8	78.9	11	8	0.3%	2	5	38	76	
Costa Rica	3,925,000	199	59.5	19	4	1.5%	2	11	25	74	
Cuba	11,290,000	264	75.5	12	7	0.5%	2	7	35	75	
Dominica	69,000	238	71.4	17	7	1.0%	2	15	28	71	
Dominican Republic	8,775,000	468	66.0	24	7	1.7%	3	34	24	66	
El Salvador	6,530,000	804	61.5	28	6	2.2%	3	27	21	67	
Guatemala	14,095,000	335	39.9	35	7	2.8%	5	38	18	64	
Haiti	7,590,000	708	36.3	34	13	2.1%	5	76	18	50	
Honduras	6,745,000	156	53.7	32	6	2.5%	4	30	19	65	
Jamaica	2,705,000	637	56.6	17	5	1.2%	2	13	27	74	
Mexico	104,340,000	138	74.6	22	5	1.7%	3	22	24	72	
Nicaragua	5,180,000	103	56.5	26	5	2.2%	3	31	20	68	
Panama	2,980,000	102	56.5	21	6	1.5%	3	21	26	70	
St. Lucia	165,000	693	38.0	21	5	1.6%	2	14	24	70	
Trinidad and Tobago	1,100,000	556	74.5	13	9	0.4%	2	25	30	67	
United States	291,680,000	77	77.4	14	8	0.6%	2	7	36	74	
SOUTH AMERICA											
Argentina	38,945,000	36	88.3	17	8	1.0%	2	16	29	72	
Bolivia	8,655,000	20	62.9	26	8	1.8%	3	56	21	62	
Brazil	183,080,000	55	81.7	18	6	1.2%	2	32	27	67	
Chile	15,745,000	54	86.1	16	6	1.0%	2	9	30	73	
Colombia	41,985,000	95	75.5	22	6	1.6%	3	22	26	67	
Ecuador	13,840,000	126	63.4	25	5	2.0%	3	32	23	69	
Guyana	705,000	9	36.7	18	9	0.9%	2	38	26	61	
Paraguay	6,115,000	39	56.7	30	5	2.6%	4	28	21	72	
Peru	28,640,000	58	73.1	23	6	1.7%	3	37	24	68	
Suriname	435,000	7	74.8	19	7	1.3%	2	25	26	67	
Uruguay	3,425,000	51	92.1	17	9	0.8%	2	14	32	73	
Venezuela	24,835,000	71	87.2	20	5	1.5%	2	24	25	71	
EUROPE											
Albania	3,535,000	318	42.9	15	5	1.0%	2	23	27	74	
Austria	8,170,000	252	67.4	9	9	0%	1	5	39	76	
Belarus	10,315,000	129	69.6	10	14	-0.4%	1	14	37	63	
Belgium	10,340,000	877	97.4	11	10	0.1%	2	5	40	75	
Bosnia and Herzegovina	4,000,000	202	43.4	13	8	0.4%	2	23	36	70	
Bulgaria	7,550,000	176	67.4	10	14	-0.5%	1	22	41	68	
Croatia	4,430,000	203	58.1	13	11	0.2%	2	7	39	71	
Czech Republic	10,250,000	337	74.5	9	11	-0.1%	1	4	38	72	
Denmark	5,405,000	325	85.1	12	11	0.1%	2	5	39	75	
Estonia	1,405,000	80	69.4	9	13	-0.4%	1	12	38	64	
Finland	5,210,000	40	58.5	11	10	0.1%	2	4	40	75	
France	60,305,000	289	75.5	13	9	0.3%	2	4	38	76	
Germany	82,415,000	598	87.7	9	10	-0.2%	1	4	41	75	
Greece	10,635,000	209	60.3	10	10	0%	1	6	40	76	
Hungary	10,045,000	280	64.8	10	13	-0.3%	1	9	38	68	
Iceland	280,000	7	92.7	14	7	0.7%	2	4	34	78	
Ireland	3,945,000	145	59.3	14	8	0.6%	2	6	33	75	
Italy	58,030,000	499	67.1	9	10	-0.1%	1	6	41	76	
Latvia	2,340,000	94	59.8	9	15	-0.6%	1	15	39	63	
Lithuania	3,590,000	142	68.6	10	13	-0.2%	1	14	37	64	
Luxembourg	460,000	460	91.9	12	8	0.4%	2	5	38	75	
Macedonia	2,065,000	208	59.4	13	8	0.5%	2	12	33	72	
Moldova	4,440,000	340	41.4	14	13	0.2%	2	42	32	61	
Netherlands	16,270,000	1,007	89.6	12	9	0.3%	2	5	39	76	
Norway	4,565,000	37	75.0	12	10	0.3%	2	4	38	77	
Poland	38,625,000	320	62.5	10	10	0.1%	1	9	36	70	
Portugal	10,110,000	285	65.8	11	10	0.1%	1	6	38	73	
Romania	22,370,000	244	55.2	11	12	-0.1%	1	28	35	67	
Serbia and Montenegro	10,660,000	270	51.7	13	11	0.2%	2	17	36	71	
Slovakia	5,420,000	286	57.6	10	10	0.1%	1	8	35	70	
Slovenia	1,935,000	247	49.1	9	10	-0.1%	1	4	39	72	
Spain	40,250,000	207	77.8	10	9	0.1%	1	5	39	76	
Sweden	8,980,000	52	83.3	11	10	0%	2	3	40	78	
Switzerland	7,430,000	466	67.3	10	8	0.1%	1	4	40	77	
Ukraine	47,890,000	205	68.0	10	16	-0.7%	1	21	38	61	
United Kingdom	60,185,000	642	89.5	11	10	0.1%	2	5	38	76	
Russia	144,310,000	22	72.9	10	14	-0.4%	1	20	38	62	
ASIA											
Afghanistan	29,205,000	116	22.3	41	17	2.3%	6	142	19	48	
Armenia	3,325,000	289	67.2	13	10	0.2%	2	41	32	62	
Azerbaijan	7,850,000	235	51.8	19	10	1.0%	2	82	27	59	
Bahrain	675,000	2,528	92.5	19	4	1.5%	3	19	29	71	
Bangladesh	139,875,000	2,516	25.6	30	9	2.1%	3	66	21	61	
Brunei	360,000	162	72.8	20	3	1.6%	2	14	26	72	
Cambodia	13,245,000	189	17.5	27	9	1.8%	4	76	19	55	
China	1,298,720,000	352	37.1	13	7	0.6%	2	25	32	70	
Cyprus	775,000	217	70.2	13	8	0.5%	2	8	34	75	
East Timor	1,010,000	176	7.5	28	6	2.1%	4	50	20	63	
Georgia	4,920,000	183	56.5	12	15	-0.3%	2	51	35	61	
India	1,057,415,000	865	27.9	23	8	1.5%	3	60	24	63	
Indonesia	236,680,000	322	42.1	21	6	1.5%	3	38	26	67	
Iran	68,650,000	108	64.7	17	6	1.2%	2	44	23	68	
Iraq	25,025,000	148	67.4	34	6	2.8%	5	55	19	67	
Israel	6,160,000	768	91.8	19	6	1.2%	3	7	29	77	
Japan	127,285,000	873	78.9	10	9	0.1%	1	3	42	78	
Jordan	5,535,000	160	78.7	24	3	2.1%	3	19	22	75	
Kazakhstan	16,780,000	16	55.8	18	11	0.8%	2	59	28	58	
Korea, North	22,585,000	485	60.5	18	7	1.1%	2	26	31	68	
Korea, South	48,450,000	1,264	82.5	13	6	0.7%	2	7	33	72	
Kuwait	2,220,000	323	96.1	22	2	1.9%	3	11	26	76	

CONTINENT/Country	Population Estimate 2004	Pop. Per Sq. Mile 2004	Percent Urban[1] 2001	Crude Birth Rate per 1,000[2] 2003	Crude Death Rate per 1,000[2] 2003	Natural Increase Percent[2] 2003	Fertility Rate (Children born/Woman)[3] 2003	Infant Mortality Rate per 1,000[3] 2003	Median Age[2] 2002	Life Expectancy Male[2] 2003	Life Expectancy Female[2] 2003
...stan	4,930,000	64	34.3	26	9	1.7%	3	75	23	59	68
...	5,995,000	66	19.7	37	12	2.5%	5	89	19	52	56
...on	3,755,000	935	90.1	20	6	1.3%	2	26	26	70	75
...sia	23,310,000	183	58.1	24	5	1.9%	3	19	24	69	75
...olia	2,730,000	5	56.6	21	7	1.4%	2	57	24	62	66
...nar	42,620,000	163	28.1	19	12	0.7%	2	70	25	54	58
...	26,770,000	471	12.2	32	10	2.3%	4	71	20	59	59
...	2,855,000	24	76.5	37	4	3.4%	6	21	19	70	75
...an	152,210,000	448	33.4	30	9	2.1%	4	77	20	61	63
...ines	85,430,000	738	59.4	26	6	2.1%	3	25	22	66	72
...	830,000	188	92.9	16	4	1.1%	3	20	31	71	76
...Arabia	24,690,000	30	86.7	37	6	3.1%	6	48	19	67	71
...ore	4,685,000	17,746	100.0	13	4	0.8%	1	4	35	77	84
...ka	19,825,000	783	23.1	16	6	1.0%	2	15	29	70	75
...	17,800,000	249	51.8	30	5	2.5%	4	32	20	68	71
...	22,675,000	1,631	(5)	13	6	0.7%	2	7	33	74	80
...tan	6,935,000	126	27.7	33	8	2.4%	4	113	19	61	68
...nd	64,570,000	326	20.0	16	7	1.0%	2	22	30	69	74
...	68,505,000	226	66.2	18	6	1.2%	2	44	27	69	74
...enistan	4,820,000	26	44.9	28	9	1.9%	4	73	21	58	65
...l Arab Emirates	2,505,000	78	87.2	18	4	1.4%	3	16	28	72	77
...istan	26,195,000	152	36.6	26	8	1.8%	3	72	22	61	68
...m	82,150,000	641	24.5	20	6	1.3%	2	31	25	68	73
...	19,680,000	97	25.0	43	9	3.4%	7	65	16	59	63
...A											
...a	33,090,000	36	57.7	22	5	1.7%	3	38	23	69	72
...a	10,875,000	23	34.9	46	26	2.0%	6	194	18	36	38
...	7,145,000	164	43.0	43	14	3.0%	6	87	16	50	52
...ana	1,570,000	7	49.4	26	31	-0.6%	3	67	19	32	32
...a Faso	13,400,000	127	16.9	45	19	2.6%	6	100	17	43	46
...di	6,165,000	574	9.3	40	18	2.2%	6	72	16	43	44
...oon	15,905,000	87	49.7	35	15	2.0%	5	70	18	47	49
...Verde	415,000	267	63.5	27	7	2.0%	4	51	19	67	73
...l African Republic	3,715,000	15	41.7	36	20	1.6%	5	93	18	40	43
...	9,395,000	19	24.1	47	16	3.1%	6	96	16	47	50
...ros	640,000	742	33.8	39	9	3.0%	5	80	19	59	64
...	2,975,000	23	66.1	29	14	1.5%	4	95	20	49	51
..., Democratic Republic of the	57,445,000	63	30.7	45	15	3.0%	7	97	16	47	51
...'Ivoire	17,145,000	138	44.0	40	18	2.2%	6	98	17	40	45
...ti	460,000	51	84.2	41	19	2.1%	6	107	18	42	44
...	75,420,000	195	42.7	24	5	1.9%	3	35	23	68	73
...orial Guinea	515,000	48	49.3	37	13	2.4%	5	89	19	53	57
...	4,390,000	97	19.1	39	13	2.6%	6	76	18	51	55
...ia	67,210,000	158	15.9	40	20	2.0%	6	103	17	40	42
...	1,340,000	13	82.3	37	11	2.5%	5	55	19	55	59
...a, The	1,525,000	370	31.3	41	12	2.8%	6	75	17	52	56
...	20,615,000	224	36.4	26	11	1.5%	3	53	20	56	57
...a	9,135,000	96	27.9	43	16	2.7%	6	93	18	48	51
...a-Bissau	1,375,000	99	32.3	38	17	2.2%	5	110	19	45	49
...	31,840,000	142	34.4	29	16	1.3%	3	63	18	45	45
...o	1,865,000	159	28.8	27	25	0.3%	4	86	20	37	37
...a	3,345,000	78	45.5	45	18	2.7%	6	132	18	47	49
...	5,565,000	8	88.0	27	3	2.4%	3	27	22	74	78
...gascar	17,235,000	76	30.1	42	12	3.0%	6	80	17	54	59
...vi	11,780,000	258	15.1	45	23	2.2%	6	105	16	38	38
...	11,790,000	25	30.9	48	19	2.9%	7	119	16	45	46
...tania	2,955,000	7	59.1	42	13	2.9%	6	74	17	50	54
...tius	1,215,000	1,542	41.6	16	7	0.9%	2	16	30	68	76
...co	31,950,000	185	56.1	23	6	1.7%	3	45	23	68	72
...mbique	18,695,000	60	33.3	37	23	1.4%	5	138	19	39	37
...ia	1,940,000	6	31.4	34	19	1.5%	5	68	18	44	41
...	11,210,000	23	21.1	50	22	2.8%	7	124	16	42	42
...a	135,570,000	380	44.9	39	14	2.5%	5	71	18	51	51
...da	7,880,000	775	6.3	40	22	1.8%	6	103	18	39	40
...ome and Principe	180,000	484	47.7	42	7	3.5%	6	46	16	65	68
...al	10,715,000	141	48.2	36	11	2.5%	5	58	18	55	58
...Leone	5,815,000	210	37.3	44	21	2.3%	6	147	18	40	45
...ia	8,165,000	33	27.9	46	18	2.9%	7	120	18	46	49
...Africa	42,770,000	91	57.7	19	18	0%	2	61	25	47	47
...	38,630,000	40	37.1	36	10	2.7%	5	66	18	57	59
...land	1,165,000	174	26.7	29	21	0.8%	4	67	19	41	38
...nia	36,230,000	99	33.3	40	17	2.2%	5	104	18	43	46
...	5,495,000	251	33.9	35	12	2.4%	5	69	17	51	55
...a	9,980,000	158	66.2	17	5	1.2%	2	27	26	73	76
...a	26,010,000	279	14.5	47	17	3.0%	7	88	15	43	46
...a	10,385,000	36	39.8	40	24	1.5%	5	99	17	35	35
...bwe	12,630,000	84	36.0	30	22	0.8%	4	66	19	40	38
...NIA											
...alia	19,825,000	7	91.2	13	7	0.5%	2	5	36	77	83
...	875,000	124	50.2	23	6	1.7%	3	13	24	66	71
...ti	100,000	319	38.6	31	9	2.3%	4	51	20	58	64
...nesia, Federated States of ...	110,000	406	28.6	26	5	2.1%	4	32	19[4]	67	71
...Zealand	3,975,000	38	85.9	14	8	0.7%	2	6	33	75	81
...New Guinea	5,360,000	30	17.6	31	8	2.3%	4	55	21	62	66
...a	180,000	165	22.3	15	6	0.9%	3	30	24	67	73
...on Islands	515,000	47	20.2	32	4	2.8%	4	23	18	70	75
...	110,000	438	33.0	25	6	1.9%	3	13	20	66	71
...tu	200,000	42	22.1	24	8	1.6%	3	58	22	60	63

...able presents data for most independent nations having an area greater than 200 square miles
...urce: United Nations World Urbanization Prospects
...urce: United States Census Bureau International Database
...urce: United States Central Intelligence Agency World Factbook
...00 Census preliminary count from www.fsmgov.org/info/people.html
...ta for Taiwan is included with China

WORLD AGRICULTURE TABLE

CONTINENT/Country	Agricultural Area 2001 Total Area Sq. Miles	Cropland Area[1] Sq. Miles	Cropland Area[1] %	Pasture Area[1] Sq. Miles	Pasture Area[1] %	Average Production 1999-2001 Wheat[1] 1,000 metric tons	Rice[1] 1,000 metric tons	Corn[1] 1,000 metric tons	Cattle[1] 1,000	Average 1999-2001 Pigs[1] 1,000	
NORTH AMERICA											
Bahamas	5,382	46	0.9%	8	0.1%	-	-	-	1	5	
Belize	8,867	402	4.5%	193	2.2%	-	12	36	52	25	
Canada	3,855,103	177,144	4.6%	111,970	2.9%	24,676	-	8,168	13,340	12,970	
Costa Rica	19,730	2,027	10.3%	9,035	45.8%	-	267	20	1,358	438	
Cuba	42,804	17,239	40.3%	8,494	19.8%	-	342	207	4,305	2,600	
Dominica	290	77	26.6%	8	2.7%	-	-	-	13	5	
Dominican Republic	18,730	6,162	32.9%	8,108	43.3%	-	615	30	2,026	548	
El Salvador	8,124	3,514	43.2%	3,066	37.7%	-	47	605	1,190	195	
Guatemala	42,042	7,355	17.5%	10,046	23.9%	9	46	1,057	2,500	1,417	
Haiti	10,714	4,247	39.6%	1,892	17.7%	-	111	211	1,390	934	
Honduras	43,277	5,514	12.7%	5,822	13.5%	1	9	509	1,737	474	
Jamaica	4,244	1,097	25.8%	884	20.8%	-	-	2	400	180	
Mexico	758,452	105,406	13.9%	308,882	40.7%	3,263	324	18,466	30,428	16,112	6
Nicaragua	50,054	8,382	16.7%	18,591	37.1%	-	234	374	2,008	402	
Panama	29,157	2,683	9.2%	5,927	20.3%	-	237	71	1,348	279	
St. Lucia	238	69	29.2%	8	3.2%	-	-	-	12	15	
Trinidad and Tobago	1,980	471	23.8%	42	2.1%	-	13	5	36	41	
United States	3,794,083	684,401	18.0%	903,479	23.8%	58,862	9,222	244,296	98,197	60,229	7
SOUTH AMERICA											
Argentina	1,073,519	135,136	12.6%	548,265	51.1%	15,642	1,140	15,217	49,299	4,200	13
Bolivia	424,165	11,973	2.8%	130,618	30.8%	121	281	607	6,715	2,786	8
Brazil	3,300,172	256,623	7.8%	760,621	23.0%	2,461	10,998	35,119	170,295	30,608	14
Chile	291,930	8,880	3.0%	49,942	17.1%	1,490	113	685	4,117	2,395	4
Colombia	439,737	16,405	3.7%	161,391	36.7%	37	2,262	1,128	25,274	2,726	2
Ecuador	109,484	11,525	10.5%	19,653	18.0%	19	1,340	483	5,261	2,654	2
Guyana	83,000	1,969	2.4%	4,749	5.7%	-	560	3	220	20	
Paraguay	157,048	12,008	7.6%	83,784	53.3%	256	112	804	9,758	2,633	
Peru	496,225	16,255	3.3%	104,634	21.1%	180	1,963	1,205	4,936	2,795	14
Suriname	63,037	259	0.4%	81	0.1%	-	178	-	128	22	
Uruguay	67,574	5,174	7.7%	52,290	77.4%	284	1,189	190	10,446	375	13
Venezuela	352,145	13,158	3.7%	70,425	20.0%	1	696	1,547	14,620	5,555	
EUROPE											
Albania	11,100	2,699	24.3%	1,699	15.3%	298	-	203	719	96	1
Austria	32,378	5,676	17.5%	7,413	22.9%	1,412	-	1,774	2,166	3,556	
Belarus	80,155	24,151	30.1%	11,564	14.4%	903	-	13	4,411	3,565	
Belgium	11,787	3,344[2]	26.2%[2]	2,618[2]	20.5%[2]	1,535	-	420	3,165	7,462	
Bosnia and Herzegovina	19,767	3,243	16.4%	4,633	23.4%	289	-	656	448	345	
Bulgaria	42,855	17,900	41.8%	6,236	14.6%	3,071	8	1,137	664	1,459	2
Croatia	21,829	6,124	28.1%	6,035	27.6%	852	-	1,958	435	1,276	
Czech Republic	30,450	12,788	42.0%	3,730	12.2%	4,196	-	324	1,604	3,761	
Denmark	16,640	8,880	53.4%	1,452	8.7%	4,683	-	-	1,887	12,052	
Estonia	17,462	2,691	15.4%	745	4.3%	123	-	-	276	304	
Finland	130,559	8,490	6.5%	77	0.1%	427	-	-	1,060	1,303	
France	208,482	75,618	36.3%	38,788	18.6%	35,327	110	15,928	20,377	14,693	9
Germany	137,847	46,409	33.7%	19,355	14.0%	21,358	-	3,362	14,723	26,021	2
Greece	50,949	14,873	29.2%	17,954	35.2%	2,111	153	2,007	584	925	8
Hungary	35,919	18,548	51.6%	4,097	11.4%	3,843	9	6,664	845	5,216	
Iceland	39,769	27	0.1%	8,780	22.1%	-	-	-	72	44	
Ireland	27,133	4,050	14.9%	12,934	47.7%	688	-	-	6,613	1,765	5
Italy	116,342	42,379	36.4%	16,907	14.5%	7,239	1,310	10,222	7,167	8,356	11
Latvia	24,942	7,220	28.9%	2,355	9.4%	410	-	-	393	407	
Lithuania	25,213	11,541	45.8%	1,923	7.6%	1,062	-	-	856	984	
Luxembourg	999	[3]	[3]	[3]	[3]	-	-	2	134	-	
Macedonia	9,928	2,363	23.8%	2,432	24.5%	308	20	135	267	209	1
Moldova	13,070	8,398	64.3%	1,483	11.3%	902	-	1,096	423	646	
Netherlands	16,164	3,622	22.4%	3,834	23.7%	995	-	148	4,108	13,253	1
Norway	125,050	3,398	2.7%	625	0.5%	265	-	-	1,017	414	2
Poland	120,728	55,267	45.8%	15,745	13.0%	8,946	-	962	6,124	17,588	
Portugal	35,516	10,444	29.4%	5,548	15.6%	295	146	907	1,415	2,346	4
Romania	91,699	38,305	41.8%	19,039	20.8%	5,610	3	8,317	3,021	5,946	8
Serbia and Montenegro	39,449	14,394	36.5%	7,197	18.2%	2,207	-	5,013	1,550	4,012	1
Slovakia	18,924	6,085	32.2%	3,375	17.8%	1,445	-	612	671	1,548	
Slovenia	7,821	784	10.0%	1,185	15.2%	153	-	283	473	585	
Spain	194,885	69,298	35.6%	44,209	22.7%	5,785	844	4,208	6,140	22,079	24
Sweden	173,732	10,413	6.0%	1,726	1.0%	2,135	-	-	1,683	1,975	4
Switzerland	15,943	1,683	10.6%	4,417	27.7%	535	-	214	1,603	1,499	4
Ukraine	233,090	129,321	55.5%	30,541	13.1%	15,043	74	3,075	10,591	9,270	1
United Kingdom	93,788	22,019	23.5%	43,440	46.3%	14,380	-	-	11,052	6,537	41
Russia	6,592,849	485,400	7.4%	351,905	5.3%	37,455	509	1,133	27,936	17,076	12
ASIA											
Afghanistan	251,773	31,097	12.4%	115,831	46.0%	1,821	205	172	2,600	-	12
Armenia	11,506	2,162	18.8%	3,089	26.8%	211	-	9	478	75	5
Azerbaijan	33,437	7,471	22.3%	10,039	30.0%	1,172	19	107	1,965	21	5
Bahrain	267	23	8.7%	15	5.8%	-	-	-	12	-	
Bangladesh	55,598	32,761	58.9%	2,317	4.2%	1,807	36,909	8	23,817	-	1
Brunei	2,226	27	1.2%	23	1.0%	-	-	-	2	6	
Cambodia	69,898	14,699	21.0%	5,792	8.3%	-	4,035	146	2,896	2,079	
China	3,690,045	599,520[4]	16.2%[4]	1,544,412[4]	41.9%[4]	102,463[4]	189,840[4]	116,240[4]	104,179[4]	440,384[4]	130
Cyprus	3,572	436	12.2%	15	0.4%	12	-	-	55	419	2
East Timor	5,743	309	5.4%	579	10.1%	-	33	93	173	300	
Georgia	26,911	4,104	15.3%	7,490	27.8%	207	-	358	1,117	433	5
India	1,222,510	655,987	53.7%	42,124	3.4%	72,140	132,818	12,285	217,773	17,000	57
Indonesia	735,310	129,730	17.6%	43,155	5.9%	-	50,953	9,409	11,370	6,098	7
Iran	636,372	63,892	10.0%	169,885	26.7%	8,740	2,103	1,113	8,273	-	53
Iraq	169,235	23,514	13.9%	15,444	9.1%	667	110	73	1,342	-	6
Israel	8,019	1,637	20.4%	548	6.8%	94	-	73	393	138	
Japan	145,850	18,510	12.7%	1,564	1.1%	657	11,551	-	4,592	9,823	3
Jordan	34,495	1,544	4.5%	2,865	8.3%	18	-	13	66	-	1
Kazakhstan	1,049,156	83,672	8.0%	714,667	68.1%	10,938	225	256	4,021	984	8
Korea, North	46,540	10,811	23.2%	193	0.4%	88	2,031	1,253	575	3,076	1
Korea, South	38,328	7,293	19.0%	208	0.5%	4	7,204	67	2,191	8,266	
Kuwait	6,880	58	0.8%	525	7.6%	-	-	-	19	-	5

	Agricultural Area 2001					Average Production 1999-2001			Average 1999-2001		
ENT/Country	Total Area Sq. Miles	Cropland Area[1] Sq. Miles	Cropland Area[1] %	Pasture Area[1] Sq. Miles	Pasture Area[1] %	Wheat[1] 1,000 metric tons	Rice[1] 1,000 metric tons	Corn[1] 1,000 metric tons	Cattle[1] 1,000	Pigs[1] 1,000	Sheep[1] 1,000
tan	77,182	5,664	7.3%	35,873	46.5%	1,113	17	363	942	98	3,101
....	91,429	3,699	4.0%	3,390	3.7%	-	2,213	108	1,106	1,390	-
....	4,016	1,208	30.1%	62	1.5%	60	-	4	76	63	354
a	127,320	29,286	23.0%	1,100	0.9%	-	2,170	63	744	1,943	167
a	604,829	4,633	0.8%	499,230	82.5%	148	-	-	2,997	17	14,587
ır	261,228	41,023	15.7%	1,212	0.5%	105	20,683	413	10,974	3,923	390
....	56,827	12,324	21.7%	6,784	11.9%	1,143	4,137	1,528	7,012	872	852
....	119,499	313	0.3%	3,861	3.2%	1	-	-	299	-	342
....	339,732	85,560	25.2%	19,305	5.7%	19,319	6,920	1,653	22,007	-	24,067
les	115,831	41,120	35.5%	4,942	4.3%	-	12,377	4,540	2,467	10,724	30
....	4,412	81	1.8%	193	4.4%	-	-	1	15	-	214
rabia	830,000	14,649	1.8%	656,373	79.1%	1,871	-	5	304	-	7,848
re	264	4	1.5%	-	0.0%	-	-	-	-	190	-
a	25,332	7,378	29.1%	1,699	6.7%	-	2,804	30	1,580	71	12
....	71,498	21,043	29.4%	31,942	44.7%	3,514	(5)	196	933	(5)	13,288
....	13,901	(5)		(5)	(5)	(5)			(5)		(5)
n	55,251	4,093	7.4%	13,514	24.5%	375	67	38	1,045	1	1,481
I	198,115	70,657	35.7%	3,089	1.6%	1	25,578	4,405	4,973	6,539	40
....	302,541	101,757	33.6%	47,792	15.8%	19,341	350	2,266	10,949	4	29,394
istan	188,457	7,008	3.7%	118,533	62.9%	1,472	33	9	863	46	5,750
Arab Emirates	32,278	919	2.8%	1,178	3.6%	-	-	-	94	-	504
tan	172,742	18,649	10.8%	88,031	51.0%	3,637	219	133	5,279	83	7,980
....	128,066	32,579	25.4%	2,479	1.9%	-	31,964	1,961	4,029	20,273	-
....	203,850	6,158	3.0%	62,027	30.4%	145	-	48	1,320	-	4,758
....	919,595	31,861	3.5%	122,780	13.4%	1,414	-	1	1,667	6	19,000
....	481,354	12,741	2.6%	208,495	43.3%	4	16	417	3,995	800	345
a	43,484	8,745	20.1%	2,124	4.9%	-	46	740	1,486	463	650
na	224,607	1,440	0.6%	98,842	44.0%	1	-	8	2,035	6	347
Faso	105,869	15,444	14.6%	23,166	21.9%	-	102	500	4,767	621	6,722
....	10,745	4,865	45.3%	3,610	33.6%	7	57	124	321	67	215
on	183,568	27,645	15.1%	7,722	4.2%	-	69	759	5,761	1,232	3,734
rde	1,557	158	10.2%	97	6.2%	-	-	27	22	195	9
African Republic	240,536	7,799	3.2%	12,066	5.0%	-	23	101	3,096	669	218
....	495,755	14,016	2.8%	173,746	35.0%	3	114	88	5,852	22	2,374
s	863	510	59.1%	58	6.7%	-	17	4	51	-	21
....	132,047	849	0.6%	38,610	29.2%	-	1	6	87	46	102
Democratic Republic of the	905,446	30,425	3.4%	57,915	6.4%	9	338	1,184	823	1,050	925
voire	124,504	28,958	23.3%	50,193	40.3%	-	1,217	693	1,398	333	1,439
....	8,958	4	0.0%	5,019	56.0%	-	-	-	269	-	465
....	386,662	12,888	3.3%	-	0.0%	6,388	5,681	6,487	3,583	29	4,510
ial Guinea	10,831	888	8.2%	402	3.7%	-	-	-	5	6	37
....	45,406	1,942	4.3%	26,900	59.2%	32	-	13	2,150	-	1,570
a	426,373	44,255	10.4%	77,220	18.1%	1,340	-	2,938	35,025	25	22,333
....	103,347	1,911	1.8%	18,012	17.4%	-	1	26	36	213	197
, The	4,127	985	23.9%	1,772	42.9%	-	28	24	350	12	115
....	92,098	22,780	24.7%	32,240	35.0%	-	244	988	1,297	327	2,715
....	94,926	5,888	6.2%	41,313	43.5%	-	830	96	2,576	93	824
Bissau	13,948	2,116	15.2%	4,170	29.9%	-	95	26	509	347	283
....	224,961	19,923	8.9%	82,240	36.6%	184	58	2,419	13,229	311	7,000
....	11,720	1,290	11.0%	7,722	65.9%	39	-	128	547	63	839
....	43,000	2,317	5.4%	7,722	18.0%	-	188	-	36	127	210
....	679,362	8,301	1.2%	51,352	7.6%	128	-	-	207	-	5,100
ıscar	226,658	13,707	6.0%	92,664	40.9%	9	2,412	175	10,339	1,267	793
....	45,747	9,035	19.7%	7,143	15.6%	2	86	2,190	741	450	110
....	478,841	18,147	3.8%	115,831	24.2%	8	801	378	6,594	72	6,282
nia	397,956	1,931	0.5%	151,545	38.1%	-	65	7	1,470	-	7,437
us	788	409	51.9%	27	3.4%	-	-	-	27	12	10
o	172,414	37,529	21.8%	81,081	47.0%	2,284	33	95	2,629	8	17,059
bique	309,496	16,351	5.3%	169,885	54.9%	1	168	1,136	1,317	179	125
a	317,818	3,166	1.0%	146,719	46.2%	4	-	26	2,436	21	2,330
....	489,192	17,375	3.6%	46,332	9.5%	10	66	5	2,217	39	4,386
a	356,669	120,464	33.8%	151,352	42.4%	75	3,109	4,734	19,677	5,000	20,833
....	10,169	5,019	49.4%	2,124	20.9%	6	13	66	766	172	264
ne and Principe	372	205	55.0%	4	1.0%	-	-	2	4	2	3
....	75,951	9,653	12.7%	21,815	28.7%	-	229	84	3,076	263	4,619
eone	27,699	2,178	7.9%	8,494	30.7%	-	215	9	413	52	365
....	246,201	4,135	1.7%	166,024	67.4%	1	2	188	5,133	4	13,100
frica	470,693	60,664	12.9%	324,048	68.8%	2,200	3	9,147	13,594	1,542	28,677
....	967,500	64,298	6.6%	452,434	46.8%	230	8	48	37,081	-	45,980
nd	6,704	734	10.9%	4,633	69.1%	-	-	94	613	32	27
a	364,900	19,112	5.2%	135,136	37.0%	87	509	2,567	17,350	449	3,513
....	21,925	10,154	46.3%	3,861	17.6%	-	69	480	277	287	1,528
....	63,170	18,954	30.0%	15,792	25.0%	1,111	-	-	760	6	6,862
....	93,065	27,799	29.9%	19,738	21.2%	12	106	1,108	5,977	1,540	1,065
....	290,586	20,386	7.0%	115,831	39.9%	80	11	768	2,709	324	137
we	150,873	12,934	8.6%	66,410	44.0%	282	-	1,698	5,840	494	602
A											
a	2,969,910	195,368	6.6%	1,563,327	52.6%	23,654	1,417	363	27,645	2,607	116,736
....	7,056	1,100	15.6%	676	9.6%	-	16	1	335	139	7
....	313	151	48.1%	-	0.0%	-	-	-	-	10	-
esia, Federated States of ...	271	139	51.3%	42	15.7%	-	-	-	14	32	-
aland	104,454	13,019	12.5%	53,525	51.2%	337	-	185	9,025	364	45,114
New Guinea	178,704	3,320	1.9%	676	0.4%	-	1	7	87	1,583	6
....	1,093	498	45.6%	8	0.7%	-	-	-	28	179	-
n Islands	10,954	286	2.6%	154	1.4%	-	5	-	11	63	-
....	251	185	73.8%	15	6.2%	-	-	-	11	81	-
ı	4,707	463	9.8%	162	3.4%	-	-	1	151	62	-

ble presents data for most independent nations having an area greater than 200 square miles
insignificant, or not available
rce: United Nations Food and Agriculture Organization
ıdes data for Luxembourg
a for Luxembourg is included with Belgium
ıdes data for Taiwan
a for Taiwan is included with China

WORLD ECONOMIC TABLE

	GDP 2002		Trade		Commercial Energy Production Avg. 2000[2]					Average Production 1999-2001 in Met			
CONTINENT/Country	Total GDP[1]	GDP Per Capita[1]	Value of Exports[1]	Value of Imports[1]	Total (1,000 Metric Tons of Coal Equiv.)	Solid %	Liquid %	Gas %	Hydro & Nuclear %	Coal[3]	Petroleum[3]	Iron Ore[4]	
NORTH AMERICA													
Bahamas	$4,590,000,000	$17,000	$560,700,000	$1,860,000,000	-	-	-	-	-	-	-	-	
Belize	$1,280,000,000	$4,900	$290,000,000	$430,000,000	12	-	-	-	100%	-	-	-	
Canada	$934,100,000,000	$29,400	$260,500,000,000	$229,000,000,000	507,218	10%	33%	43%	14%	70,711,084	97,834,913	20,527,000	
Costa Rica	$32,000,000,000	$8,500	$5,100,000,000	$6,400,000,000	1,937	-	-	-	100%	-	-	-	
Cuba	$30,690,000,000	$2,300	$1,800,000,000	$4,800,000,000	4,626	-	83%	17%	-	-	2,134,520	-	
Dominica	$380,000,000	$5,400	$50,000,000	$135,000,000	4	-	-	-	100%	-	-	-	
Dominican Republic	$53,780,000,000	$6,100	$5,300,000,000	$8,700,000,000	115	-	-	-	100%	-	-	-	
El Salvador	$29,410,000,000	$4,700	$3,000,000,000	$4,900,000,000	1,110	-	-	-	100%	-	-	-	
Guatemala	$53,200,000,000	$3,700	$2,700,000,000	$5,600,000,000	1,822	-	81%	1%	18%	-	1,076,526	9,000	
Haiti	$10,600,000,000	$1,700	$298,000,000	$1,140,000,000	33	-	-	-	100%	-	-	-	
Honduras	$16,290,000,000	$2,600	$1,300,000,000	$2,700,000,000	347	-	-	-	100%	-	-	-	
Jamaica	$10,080,000,000	$3,900	$1,400,000,000	$3,100,000,000	18	-	-	-	100%	-	-	-	11,
Mexico	$924,400,000,000	$9,000	$158,400,000,000	$168,400,000,000	340,594	1%	79%	16%	4%	11,097,943	150,165,451	6,860,000	
Nicaragua	$11,160,000,000	$2,500	$637,000,000	$1,700,000,000	706	-	-	-	100%	-	-	-	
Panama	$18,060,000,000	$6,000	$5,800,000,000	$6,700,000,000	418	-	-	-	100%	-	-	-	
St. Lucia	$866,000,000	$5,400	$68,300,000	$319,400,000	-	-	-	-	-	-	-	-	
Trinidad and Tobago	$11,070,000,000	$9,500	$4,200,000,000	$3,800,000,000	22,768	-	39%	61%	-	-	5,964,991	-	
United States	$10,450,000,000,000	$37,600	$733,900,000,000	$1,194,100,000,000	2,342,228	33%	22%	30%	14%	996,498,186	289,640,487	35,178,000	
SOUTH AMERICA													
Argentina	$403,800,000,000	$10,200	$25,300,000,000	$9,000,000,000	118,739	-	50%	45%	5%	260,299	38,783,798	-	
Bolivia	$21,150,000,000	$2,500	$1,300,000,000	$1,600,000,000	7,732	-	33%	64%	3%	-	1,599,401	-	
Brazil	$1,376,000,000,000	$7,600	$59,400,000,000	$46,200,000,000	143,640	3%	63%	6%	28%	4,446,477	61,155,586	124,667,000	13,
Chile	$156,100,000,000	$10,000	$17,800,000,000	$15,600,000,000	6,180	6%	11%	45%	38%	475,484	349,201	5,523,000	
Colombia	$251,600,000,000	$6,500	$12,900,000,000	$12,500,000,000	99,513	36%	52%	9%	4%	38,112,136	34,896,672	348,000	
Ecuador	$42,650,000,000	$3,100	$4,900,000,000	$6,000,000,000	32,171	-	94%	3%	3%	-	19,520,185	-	
Guyana	$2,628,000,000	$4,000	$500,000,000	$575,000,000	1	-	-	-	100%	-	-	-	2,2
Paraguay	$25,190,000,000	$4,200	$2,000,000,000	$2,400,000,000	6,577	-	-	-	100%	-	-	-	
Peru	$138,800,000,000	$4,800	$7,600,000,000	$7,300,000,000	10,933	-	73%	9%	18%	52,297	4,932,561	2,701,000	
Suriname	$1,469,000,000	$3,500	$445,000,000	$300,000,000	1,022	-	84%	-	16%	-	496,400	-	3,9
Uruguay	$26,820,000,000	$7,800	$2,100,000,000	$1,870,000,000	867	-	-	-	100%	-	-	-	
Venezuela	$131,700,000,000	$5,500	$28,600,000,000	$18,800,000,000	311,899	3%	81%	14%	2%	7,482,998	146,621,238	10,497,000	4,3
EUROPE													
Albania	$15,690,000,000	$4,500	$340,000,000	$1,500,000,000	1,089	1%	42%	2%	55%	32,666	284,321	-	
Austria	$227,700,000,000	$27,700	$70,000,000,000	$74,000,000,000	9,611	5%	15%	24%	56%	1,197,660	921,120	525,000	
Belarus	$90,190,000,000	$8,200	$7,700,000,000	$8,800,000,000	3,644	18%	73%	9%	-	-	1,830,872	-	
Belgium	$299,700,000,000	$29,000	$162,000,000,000	$152,000,000,000	18,451	2%	-	-	98%	318,998	-	-	
Bosnia and Herzegovina	$7,300,000,000	$1,900	$1,150,000,000	$2,800,000,000	6,553	90%	-	-	10%	8,414,623	-	50,000	
Bulgaria	$49,230,000,000	$6,600	$5,300,000,000	$6,900,000,000	13,500	46%	-	-	53%	28,841,963	37,048	310,000	
Croatia	$43,120,000,000	$8,800	$4,900,000,000	$10,700,000,000	4,962	-	42%	43%	15%	5,104	1,191,360	-	
Czech Republic	$157,100,000,000	$15,300	$40,800,000,000	$43,200,000,000	39,843	85%	1%	1%	14%	63,466,671	283,097	-	
Denmark	$155,300,000,000	$29,000	$56,300,000,000	$47,900,000,000	36,502	-	70%	29%	2%	-	16,701,163	-	
Estonia	$15,520,000,000	$10,900	$3,400,000,000	$4,400,000,000	3,892	100%	-	-	-	-	-	-	
Finland	$133,800,000,000	$26,200	$40,100,000,000	$31,800,000,000	11,933	15%	-	-	85%	-	-	-	
France	$1,558,000,000,000	$25,700	$307,800,000,000	$303,700,000,000	175,306	2%	4%	1%	93%	3,616,981	1,446,228	12,000	
Germany	$2,160,000,000,000	$26,600	$608,000,000,000	$487,300,000,000	181,697	47%	2%	13%	38%	204,685,080	3,044,206	5,000	
Greece	$203,300,000,000	$19,000	$12,600,000,000	$31,400,000,000	12,988	92%	3%	1%	4%	64,503,999	166,807	583,000	1,9
Hungary	$134,000,000,000	$13,300	$31,400,000,000	$33,900,000,000	16,319	25%	19%	24%	32%	14,796,257	1,301,710	-	9
Iceland	$8,444,000,000	$25,000	$2,300,000,000	$2,100,000,000	1,638	-	-	-	100%	-	-	-	
Ireland	$113,700,000,000	$30,500	$86,600,000,000	$48,600,000,000	3,232	47%	-	47%	6%	-	-	-	
Italy	$1,455,000,000,000	$25,000	$259,200,000,000	$238,200,000,000	40,332	-	16%	54%	30%	47,666	4,144,278	-	
Latvia	$20,990,000,000	$8,300	$2,300,000,000	$3,900,000,000	369	6%	-	-	94%	-	-	-	
Lithuania	$30,080,000,000	$8,400	$5,400,000,000	$6,800,000,000	3,677	-	12%	-	87%	-	251,824	-	
Luxembourg	$21,940,000,000	$44,000	$10,100,000,000	$13,250,000,000	113	-	-	-	100%	-	-	-	
Macedonia	$10,570,000,000	$5,000	$1,100,000,000	$1,900,000,000	3,038	95%	-	-	5%	7,463,628	-	9,000	
Moldova	$11,510,000,000	$2,500	$590,000,000	$980,000,000	7	-	-	-	100%	-	-	-	
Netherlands	$437,800,000,000	$26,900	$243,300,000,000	$201,100,000,000	87,974	-	4%	94%	2%	-	1,437,293	-	
Norway	$149,100,000,000	$31,800	$68,200,000,000	$37,300,000,000	324,396	-	72%	22%	5%	847,996	154,419,533	355,000	
Poland	$373,200,000,000	$9,500	$32,400,000,000	$43,400,000,000	108,277	94%	1%	5%	-	164,737,813	645,072	-	
Portugal	$195,200,000,000	$18,000	$25,900,000,000	$39,000,000,000	1,560	-	-	-	100%	-	-	6,000	
Romania	$169,300,000,000	$7,400	$13,700,000,000	$16,700,000,000	37,598	19%	24%	46%	10%	27,392,191	6,038,110	24,000	
Serbia and Montenegro	$23,150,000,000	$2,370	$2,400,000,000	$6,300,000,000	14,188	74%	8%	8%	10%	34,480,488	810,787	10,000	5
Slovakia	$67,340,000,000	$12,200	$12,900,000,000	$15,400,000,000	8,813	17%	1%	2%	79%	3,606,648	48,134	200,000	
Slovenia	$37,060,000,000	$18,000	$10,300,000,000	$11,100,000,000	3,644	38%	-	-	62%	4,391,644	991	-	
Spain	$850,700,000,000	$20,700	$122,200,000,000	$156,600,000,000	40,444	28%	2%	1%	68%	23,479,212	296,665	-	
Sweden	$230,700,000,000	$25,400	$80,600,000,000	$68,600,000,000	31,413	1%	-	-	99%	-	-	12,114,000	
Switzerland	$233,400,000,000	$31,700	$100,300,000,000	$94,400,000,000	14,710	-	-	-	100%	-	-	-	
Ukraine	$218,000,000,000	$4,500	$18,100,000,000	$18,000,000,000	118,973	50%	5%	20%	25%	81,998,575	3,747,936	28,933,000	
United Kingdom	$1,528,000,000,000	$25,300	$286,300,000,000	$330,100,000,000	397,906	7%	47%	38%	8%	32,758,497	119,820,635	1,000	
Russia	$1,409,000,000,000	$9,300	$104,600,000,000	$60,700,000,000	1,412,286	10%	33%	52%	5%	253,376,954	324,436,632	48,300,000	3,9
ASIA													
Afghanistan	$19,000,000,000	$700	$1,200,000,000	$1,300,000,000	195	1%	-	79%	20%	1,000	-	-	
Armenia	$12,130,000,000	$3,800	$525,000,000	$991,000,000	901	-	-	-	100%	-	-	-	
Azerbaijan	$28,610,000,000	$3,500	$2,000,000,000	$1,800,000,000	27,748	-	72%	27%	1%	-	14,183,985	-	
Bahrain	$9,910,000,000	$14,000	$5,800,000,000	$4,200,000,000	14,442	-	22%	78%	-	-	1,827,397	-	
Bangladesh	$238,200,000,000	$1,700	$6,200,000,000	$8,500,000,000	11,713	-	-	99%	1%	-	120,476	-	
Brunei	$6,500,000,000	$18,600	$3,000,000,000	$1,400,000,000	27,922	-	49%	51%	-	-	9,435,323	-	
Cambodia	$20,420,000,000	$1,500	$1,380,000,000	$1,730,000,000	10	-	-	-	100%	-	-	-	
China	$5,989,000,000,000	$4,400	$658,260,000,000	$618,930,000,000	1,023,314[5]	70%[5]	23%[5]	4%[5]	3%[5]	1,251,423,183	161,226,848	72,967,000	9,0
Cyprus	$9,400,000,000	$15,000	$1,030,000,000	$3,900,000,000	-	-	-	-	-	-	-	-	
East Timor	$440,000,000	$500	$8,000,000	$237,000,000	-	-	-	-	-	-	-	-	
Georgia	$16,050,000,000	$3,100	$515,000,000	$750,000,000	963	1%	16%	8%	75%	10,000	102,258	-	
India	$2,664,000,000,000	$2,540	$44,500,000,000	$53,800,000,000	367,807	73%	14%	8%	4%	304,842,421	32,123,682	48,080,000	7,5
Indonesia	$714,200,000,000	$3,100	$52,300,000,000	$32,100,000,000	279,695	27%	45%	26%	2%	79,664,587	70,565,213	282,000	1,1
Iran	$458,300,000,000	$7,000	$24,800,000,000	$21,800,000,000	350,729	-	77%	23%	-	1,376,993	181,632,777	5,367,000	1
Iraq	$58,000,000,000	$2,400	$13,000,000,000	$7,800,000,000	186,519	-	97%	3%	-	-	124,281,583	-	
Israel	$117,400,000,000	$19,000	$28,100,000,000	$30,800,000,000	334	94%	2%	4%	1%	-	5,957	-	
Japan	$3,651,000,000,000	$28,000	$383,800,000,000	$292,100,000,000	142,731	2%	1%	2%	95%	3,286,983	351,650	1,000	
Jordan	$22,630,000,000	$4,300	$2,500,000,000	$4,400,000,000	316	-	1%	97%	2%	-	1,986	-	
Kazakhstan	$120,000,000,000	$6,300	$10,300,000,000	$9,600,000,000	113,390	40%	45%	14%	1%	70,311,969	30,508,827	7,467,000	3,6
Korea, North	$22,260,000,000	$1,000	$842,000,000	$1,314,000,000	65,932	96%	-	-	4%	94,174,845	-	3,000,000	
Korea, South	$941,500,000,000	$19,400	$162,600,000,000	$148,400,000,000	43,892	6%	-	-	94%	4,054,646	-	175,000	
Kuwait	$36,850,000,000	$15,000	$16,000,000,000	$7,300,000,000	161,322	-	92%	8%	-	-	98,844,823	-	

NT/Country	GDP 2002 Total GDP[1]	GDP Per Capita[1]	Trade Value of Exports[1]	Value of Imports[1]	Commercial Energy Production Avg. 2000[2] Total (1,000 Metric Tons of Coal Equiv.)	Solid %	Liquid %	Gas %	Hydro & Nuclear %	Average Production 1999-2001 in Metric Tons Coal[3]	Petroleum[3]	Iron Ore[4]	Bauxite[4]
an	$13,880,000,000	$2,800	$488,000,000	$587,000,000	2,026	9%	5%	2%	83%	423,664	91,503	-	-
	$10,400,000,000	$1,700	$345,000,000	$555,000,000	146	1%	-	-	99%	1,000	-	-	-
	$17,610,000,000	$5,400	$1,000,000,000	$6,000,000,000	55	-	-	-	100%	-	-	-	-
	$198,400,000,000	$9,300	$95,200,000,000	$76,800,000,000	110,069	-	41%	58%	1%	314,332	33,792,132	208,000	137,000
a	$5,060,000,000	$1,840	$501,000,000	$659,000,000	2,212	100%	-	-	-	5,099,640	-	-	-
r	$73,690,000,000	$1,660	$2,700,000,000	$2,500,000,000	9,297	3%	6%	88%	2%	358,331	587,374	-	-
	$37,320,000,000	$1,400	$720,000,000	$1,600,000,000	172	10%	-	-	90%	9,667	-	-	-
	$22,400,000,000	$8,300	$10,600,000,000	$5,500,000,000	74,376	-	92%	8%	-	-	46,989,489	-	-
	$295,300,000,000	$2,100	$9,800,000,000	$11,100,000,000	33,773	6%	12%	74%	7%	3,247,391	2,768,108	-	10,000
es	$379,700,000,000	$4,200	$35,100,000,000	$33,500,000,000	16,244	6%	-	-	94%	1,306,993	173,128	-	-
	$15,910,000,000	$21,500	$10,900,000,000	$3,900,000,000	92,237	-	57%	43%	-	-	35,018,538	-	-
abia	$268,900,000,000	$10,500	$71,000,000,000	$39,500,000,000	736,996	-	91%	9%	-	-	401,559,222	-	-
re	$112,400,000,000	$24,000	$127,000,000,000	$113,000,000,000	-	-	-	-	-	-	-	-	-
a	$73,700,000,000	$3,700	$4,600,000,000	$5,400,000,000	394	-	-	-	100%	-	-	-	-
	$63,480,000,000	$3,500	$6,200,000,000	$4,900,000,000	47,898	-	83%	15%	2%	-	26,119,029	-	-
	$406,000,000,000	$18,000	$130,000,000,000	$113,000,000,000	[6]	[6]	[6]	[6]	[6]	58,284	38,686	-	-
n	$8,476,000,000	$1,250	$710,000,000	$830,000,000	1,790	-	1%	3%	95%	20,667	16,613	-	-
	$445,800,000,000	$6,900	$67,700,000,000	$58,100,000,000	44,127	25%	24%	50%	2%	18,551,756	5,080,720	20,000	-
	$489,700,000,000	$7,000	$35,100,000,000	$50,800,000,000	28,167	69%	14%	3%	14%	65,334,995	2,642,106	2,300,000	303,000
istan	$31,340,000,000	$5,500	$2,970,000,000	$2,250,000,000	71,764	-	15%	85%	-	-	7,139,688	-	-
Arab Emirates	$53,970,000,000	$22,000	$44,900,000,000	$30,800,000,000	199,656	-	83%	17%	-	-	112,737,023	-	-
an	$66,060,000,000	$2,500	$2,800,000,000	$2,500,000,000	85,806	1%	13%	85%	1%	2,736,319	4,419,300	-	-
	$183,800,000,000	$2,250	$16,500,000,000	$16,800,000,000	39,300	30%	59%	5%	7%	9,688,950	15,926,911	-	-
	$15,070,000,000	$840	$3,400,000,000	$2,900,000,000	30,622	-	100%	-	-	-	21,304,264	-	-
	$173,800,000,000	$5,300	$19,500,000,000	$10,600,000,000	222,648	-	47%	53%	-	24,000	61,651,110	757,000	-
	$18,360,000,000	$1,600	$8,600,000,000	$4,100,000,000	53,315	-	98%	1%	-	-	36,961,745	-	-
	$7,380,000,000	$1,070	$207,000,000	$479,000,000	69	-	100%	-	-	-	39,547	-	-
a	$13,480,000,000	$9,500	$2,400,000,000	$1,900,000,000	[7]	[7]	[7]	[7]	[7]	956,767	-	-	-
Faso	$14,510,000,000	$1,080	$250,000,000	$525,000,000	15	-	-	-	100%	-	-	-	-
	$3,146,000,000	$600	$26,000,000	$135,000,000	21	29%	-	-	71%	-	-	-	-
on	$26,840,000,000	$1,700	$1,900,000,000	$1,700,000,000	10,722	-	96%	-	4%	1,000	4,326,440	-	-
rde	$600,000,000	$1,400	$30,000,000	$220,000,000	-	-	-	-	-	-	-	-	-
African Republic	$4,296,000,000	$1,300	$134,000,000	$102,000,000	10	-	-	-	100%	-	-	-	-
	$9,297,000,000	$1,100	$197,000,000	$570,000,000	-	-	-	-	-	-	-	-	-
s	$441,000,000	$720	$16,300,000	$39,800,000	-	-	-	-	-	-	-	-	-
	$2,500,000,000	$900	$2,400,000,000	$73,000,000	19,097	-	99%	1%	-	-	13,651,000	-	-
Democratic Republic of the	$34,000,000,000	$610	$1,200,000,000	$890,000,000	2,630	4%	71%	-	25%	96,000	1,194,669	-	-
voire	$24,030,000,000	$1,500	$4,400,000,000	$2,500,000,000	4,439	-	50%	45%	5%	-	620,450	-	-
	$619,000,000	$1,300	$70,000,000	$255,000,000	-	-	-	-	-	-	-	-	-
	$289,800,000,000	$3,900	$7,000,000,000	$15,200,000,000	86,315	-	65%	32%	2%	-	38,024,058	1,283,000	-
ial Guinea	$1,270,000,000	$2,700	$2,500,000,000	$562,000,000	7,531	-	100%	-	-	-	7,461,521	-	-
	$3,300,000,000	$740	$20,000,000	$500,000,000	-	-	-	-	-	-	-	-	-
	$48,530,000,000	$750	$433,000,000	$1,630,000,000	211	-	-	-	100%	-	-	-	-
	$8,354,000,000	$5,700	$2,600,000,000	$1,100,000,000	23,273	-	95%	5%	-	-	15,674,359	-	-
The	$2,582,000,000	$1,800	$138,000,000	$225,000,000	-	-	-	-	-	-	-	-	-
	$41,250,000,000	$2,100	$2,200,000,000	$2,800,000,000	830	-	2%	-	98%	-	330,933	-	525,000
	$18,690,000,000	$2,000	$835,000,000	$670,000,000	25	-	-	-	100%	-	-	-	15,663,000
Bissau	$901,400,000	$800	$71,000,000	$59,000,000	-	-	-	-	-	-	-	-	-
	$32,890,000,000	$1,020	$2,100,000,000	$3,000,000,000	642	-	-	-	100%	-	-	-	-
	$5,106,000,000	$2,700	$422,000,000	$738,000,000	[7]	[7]	[7]	[7]	[7]	-	-	-	-
	$3,116,000,000	$1,100	$110,000,000	$165,000,000	24	-	-	-	100%	-	-	-	-
	$33,360,000,000	$7,600	$11,800,000,000	$6,300,000,000	103,205	-	92%	8%	-	-	67,767,436	-	-
scar	$12,590,000,000	$760	$700,000,000	$985,000,000	64	-	-	-	100%	-	-	-	-
	$6,811,000,000	$670	$435,000,000	$505,000,000	107	-	-	-	100%	-	-	-	-
	$9,775,000,000	$860	$680,000,000	$630,000,000	29	-	-	-	100%	-	-	-	-
nia	$4,891,000,000	$1,900	$355,000,000	$360,000,000	4	-	-	-	100%	-	-	7,492,000	-
us	$12,150,000,000	$11,000	$1,600,000,000	$1,800,000,000	12	-	-	-	100%	-	-	-	-
)	$121,800,000,000	$3,900	$7,500,000,000	$10,400,000,000	201	14%	9%	33%	43%	61,000	15,223	4,000	-
bique	$19,520,000,000	$1,000	$680,000,000	$1,180,000,000	874	2%	-	-	98%	18,667	-	-	8,000
	$13,150,000,000	$6,900	$1,210,000,000	$1,380,000,000	[7]	[7]	[7]	[7]	[7]	-	-	-	-
	$8,713,000,000	$830	$293,000,000	$368,000,000	175	100%	-	-	-	151,666	-	-	-
	$112,500,000,000	$875	$17,300,000,000	$13,600,000,000	172,641	-	90%	10%	-	61,000	108,397,478	-	-
	$8,920,000,000	$1,200	$68,000,000	$253,000,000	20	-	-	-	100%	-	-	-	-
e and Principe	$200,000,000	$1,200	$5,500,000	$24,800,000	1	-	-	-	100%	-	-	-	-
	$15,640,000,000	$1,500	$1,150,000,000	$1,460,000,000	1	-	-	100%	-	-	-	-	-
eone	$2,826,000,000	$580	$35,000,000	$190,000,000	[7]	[7]	[7]	[7]	[7]	-	-	-	-
	$4,270,000,000	$550	$126,000,000	$343,000,000	-	-	-	-	-	-	-	-	-
frica	$427,700,000,000	$10,000	$31,800,000,000	$26,600,000,000	245,195[8]	92%[8]	5%[8]	1%[8]	2%[8]	224,286,505	1,277,485	20,751,000	-
	$52,900,000,000	$1,420	$1,800,000,000	$1,500,000,000	13,436	-	99%	-	1%	-	7,679,837	-	-
nd	$5,542,000,000	$4,400	$820,000,000	$938,000,000	[7]	[7]	[7]	[7]	[7]	288,665	-	-	-
a	$20,420,000,000	$630	$863,000,000	$1,670,000,000	343	23%	-	-	77%	5,000	-	-	-
	$7,594,000,000	$1,500	$449,000,000	$561,000,000	-	-	-	-	-	-	-	-	-
	$67,130,000,000	$6,500	$6,800,000,000	$8,700,000,000	8,065	-	66%	34%	-	-	3,826,400	105,000	-
	$30,490,000,000	$1,260	$476,000,000	$1,140,000,000	193	-	-	-	100%	-	-	3,000	-
	$8,240,000,000	$890	$709,000,000	$1,123,000,000	1,117	15%	-	-	85%	192,358	-	-	-
we	$26,070,000,000	$2,400	$1,570,000,000	$1,739,000,000	4,801	92%	-	-	8%	4,508,643	-	237,000	-
A													
a	$525,500,000,000	$27,000	$66,300,000,000	$68,000,000,000	331,923	71%	14%	14%	1%	307,176,075	31,728,994	104,014,000	51,834,000
	$4,822,000,000	$5,500	$442,000,000	$642,000,000	53	-	-	-	100%	-	-	-	-
	$79,000,000	$840	$6,000,000	$44,000,000	-	-	-	-	-	-	-	-	-
sia, Federated States of	$277,000,000	$2,000	$22,000,000	$149,000,000	-	-	-	-	-	-	-	-	-
aland	$78,400,000,000	$20,200	$15,000,000,000	$12,500,000,000	19,812	14%	13%	40%	33%	3,452,315	1,839,394	660,000	-
lew Guinea	$10,860,000,000	$2,300	$1,800,000,000	$1,100,000,000	5,864	-	96%	2%	2%	-	3,874,601	-	-
	$1,000,000,000	$5,600	$15,500,000	$130,100,000	3	-	-	-	100%	-	-	-	-
n Islands	$800,000,000	$1,700	$47,000,000	$82,000,000	-	-	-	-	-	-	-	-	-
	$236,000,000	$2,200	$8,900,000	$70,000,000	-	-	-	-	-	-	-	-	-
ı	$563,000,000	$2,900	$22,000,000	$93,000,000	-	-	-	-	-	-	-	-	-

le presents data for most independent nations having an area greater than 200 square miles
nsignificant, or not available
ce: United States Central Intelligence Agency World Factbook
ce: United Nations Energy Statistics Yearbook
ce: United States Energy Information Administration International Energy Annual
ce: United States Geological Survey Minerals Yearbook
des data for Taiwan
for Taiwan is included with China
for countries in the South Africa Customs Union are included with South Africa
des data for countries in the South Africa Customs Union

WORLD ENVIRONMENT TABLE

CONTINENT/Country	Total Area Sq. Miles	Protected Area 2002[1,2] Sq. Miles	%	Endangered Species 2003[3] Mammal	Bird	Reptile	Amphib.	Fish	Invrt.	Forest Cover Sq. Miles 2000	Perce 19
NORTH AMERICA											
Bahamas	5,382	-	-	5	4	6	0	15	1	3,251	
Belize	8,867	3,999	45.1%	5	2	4	0	17	1	5,205	
Canada	3,855,103	427,916	11.1%	16	8	2	1	25	11	944,294	
Costa Rica	19,730	4,538	23.0%	13	13	7	1	13	9	7,598	
Cuba	42,804	29,578	69.1%	11	18	7	0	23	3	9,066	
Dominica	290	-	-	1	3	4	0	11	0	178	
Dominican Republic	18,730	9,721	51.9%	5	15	10	1	10	2	5,313	
El Salvador	8,124	33	0.4%	2	0	4	0	5	1	467	
Guatemala	42,042	8,408	20.0%	7	6	8	0	14	8	11,004	
Haiti	10,714	43	0.4%	4	14	8	1	12	2	340	
Honduras	43,277	2,770	6.4%	10	5	6	0	14	2	20,784	
Jamaica	4,244	3,590	84.6%	5	12	8	4	12	5	1,255	
Mexico	758,452	77,362	10.2%	72	40	18	4	106	41	213,148	
Nicaragua	50,054	8,910	17.8%	6	5	7	0	17	2	12,656	
Panama	29,157	6,327	21.7%	17	16	7	0	17	2	11,104	
St. Lucia	238	-	-	2	5	6	0	10	0	35	
Trinidad and Tobago	1,980	119	6.0%	1	1	5	0	15	0	1,000	
United States	3,794,083	982,668	25.9%	39	56	27	25	155	557	872,563	
SOUTH AMERICA											
Argentina	1,073,519	70,852	6.6%	32	39	5	5	9	10	133,777	
Bolivia	424,165	56,838	13.4%	25	28	2	1	0	1	204,897	
Brazil	3,300,172	221,112	6.7%	74	113	22	6	33	34	2,100,028	
Chile	291,930	55,175	18.9%	21	22	0	3	9	0	59,985	
Colombia	439,737	44,853	10.2%	39	78	14	0	23	0	191,510	
Ecuador	109,484	20,036	18.3%	34	62	10	0	11	48	40,761	
Guyana	83,000	249	0.3%	13	2	6	0	13	1	65,170	
Paraguay	157,048	5,497	3.5%	10	26	2	0	0	0	90,240	
Peru	496,225	30,270	6.1%	46	76	6	1	8	2	251,796	
Suriname	63,037	3,089	4.9%	12	1	6	0	12	0	54,491	
Uruguay	67,574	203	0.3%	6	11	3	0	8	1	4,988	
Venezuela	352,145	224,669	63.8%	26	24	13	0	19	1	191,144	
EUROPE											
Albania	11,100	422	3.8%	3	3	4	0	16	4	3,826	
Austria	32,378	10,685	33.0%	7	3	0	0	7	44	15,004	
Belarus	80,155	5,050	6.3%	7	3	0	0	0	5	36,301	
Belgium	11,787	-	-	11	2	0	0	7	11	2,811	
Bosnia and Herzegovina	19,767	99	0.5%	10	3	1	1	10	10	8,776	
Bulgaria	42,855	1,928	4.5%	14	10	2	0	10	9	14,247	
Croatia	21,829	1,637	7.5%	9	4	1	1	26	11	6,884	
Czech Republic	30,450	4,902	16.1%	8	2	0	0	7	19	10,162	
Denmark	16,640	5,658	34.0%	5	1	0	0	7	11	1,757	
Estonia	17,462	2,061	11.8%	5	3	0	0	1	4	7,954	
Finland	130,559	12,142	9.3%	4	3	0	0	1	10	84,691	
France	208,482	27,728	13.3%	18	5	3	2	15	65	59,232	
Germany	137,847	43,973	31.9%	11	5	0	0	12	31	41,467	
Greece	50,949	1,834	3.6%	13	7	6	1	26	11	13,896	
Hungary	35,919	2,514	7.0%	9	8	1	0	8	25	7,104	
Iceland	39,769	3,897	9.8%	7	0	0	0	8	0	120	
Ireland	27,133	461	1.7%	6	1	0	0	6	3	2,544	
Italy	116,342	9,191	7.9%	14	5	4	4	16	58	38,622	
Latvia	24,942	3,342	13.4%	5	3	0	0	3	8	11,286	
Lithuania	25,213	2,597	10.3%	6	4	0	0	3	5	7,699	
Luxembourg	999	-	-	3	1	0	0	0	4	-	
Macedonia	9,928	705	7.1%	11	3	2	0	4	5	3,498	
Moldova	13,070	183	1.4%	6	5	1	0	9	5	1,255	
Netherlands	16,164	2,295	14.2%	10	4	0	0	7	7	1,448	
Norway	125,050	8,503	6.8%	10	2	0	0	7	9	34,240	
Poland	120,728	14,970	12.4%	14	4	0	0	3	15	34,931	
Portugal	35,516	2,344	6.6%	17	7	0	1	19	82	14,154	
Romania	91,699	4,310	4.7%	17	8	2	0	10	22	24,896	
Serbia and Montenegro	39,449	1,302	3.3%	12	5	1	0	19	19	11,147	
Slovakia	18,924	4,315	22.8%	9	4	1	0	8	19	8,405	
Slovenia	7,821	469	6.0%	9	1	0	1	15	42	4,274	
Spain	194,885	16,565	8.5%	24	7	7	3	23	63	55,483	
Sweden	173,732	15,810	9.1%	6	2	0	0	6	13	104,765	
Switzerland	15,943	4,783	30.0%	5	2	0	0	4	30	4,629	
Ukraine	233,090	9,091	3.9%	16	8	2	0	11	14	37,004	
United Kingdom	93,788	19,602	20.9%	12	2	0	0	11	10	10,788	
Russia	6,592,849	514,242	7.8%	45	38	6	0	18	30	3,287,242	
ASIA											
Afghanistan	251,773	755	0.3%	13	11	1	1	0	1	5,216	
Armenia	11,506	874	7.6%	11	4	5	0	1	7	1,355	
Azerbaijan	33,437	2,040	6.1%	13	8	5	0	5	6	4,224	
Bahrain	267	-	-	1	6	0	0	6	0	-	
Bangladesh	55,598	445	0.8%	22	23	20	0	8	0	5,151	
Brunei	2,226	-	-	11	14	4	0	6	0	1,707	
Cambodia	69,898	12,931	18.5%	24	19	10	0	11	0	36,043	
China	3,690,045	287,824	7.8%	81	75	31	1	46	4	631,200	
Cyprus	3,572	-	-	3	3	3	0	6	0	664	
East Timor	5,743	-	-	0	6	0	0	2	0	1,958	
Georgia	26,911	619	2.3%	13	3	7	1	6	10	11,537	
India	1,222,510	63,571	5.2%	86	72	25	3	27	23	247,542	
Indonesia	735,310	151,474	20.6%	147	114	28	0	91	31	405,353	
Iran	636,372	30,546	4.8%	22	13	8	2	14	3	28,182	
Iraq	169,235	-	-	11	11	2	0	3	2	3,085	
Israel	8,019	1,267	15.8%	15	12	4	0	10	10	510	
Japan	145,850	9,918	6.8%	37	35	11	10	27	45	92,977	
Jordan	34,495	1,173	3.4%	9	8	1	0	5	3	332	
Kazakhstan	1,049,156	28,327	2.7%	17	15	2	1	7	4	46,904	
Korea, North	46,540	1,210	2.6%	13	19	0	0	5	1	31,699	
Korea, South	38,328	2,645	6.9%	13	25	0	0	7	1	24,124	

ENT/Country	Total Area Sq. Miles	Protected Area 2002[1,2] Sq. Miles	%	Mammal	Bird	Endangered Species 2003[3] Reptile	Amphib.	Fish	Invrt.	Forest Cover[4] Sq. Miles 2000	Percent Change 1990-2000
....	6,880	103	1.5%	1	7	1	0	6	0	19	66.7%
...stan	77,182	2,779	3.6%	7	4	2	0	0	3	3,873	29.4%
....	91,429	11,429	12.5%	31	20	11	0	6	0	48,498	-4.0%
...n	4,016	20	0.5%	6	7	1	0	8	1	139	-2.7%
...a	127,320	7,257	5.7%	50	37	21	0	34	3	74,487	52.4%
...ia	604,829	69,555	11.5%	14	16	0	0	1	3	41,101	-5.3%
...ar	261,228	784	0.3%	39	35	20	0	7	2	132,892	-13.1%
....	56,827	5,058	8.9%	29	25	6	0	0	1	15,058	-16.7%
....	119,499	16,730	14.0%	11	10	4	0	17	1	4	-
...n	339,732	16,647	4.9%	17	17	9	0	14	0	9,116	-14.3%
...nes	115,831	6,602	5.7%	50	67	8	23	48	19	22,351	-13.3%
....	4,412	-	-	0	6	1	0	4	0		
...rabia	830,000	317,890	38.3%	9	15	2	0	8	1	5,807	-
...ore	264	13	4.9%	3	7	3	0	12	1	8	-
...ka	25,332	3,420	13.5%	22	14	8	0	22	2	7,490	-15.2%
....	71,498	-	-	4	8	3	0	8	3	1,780	-
....	13,901	-	-	12	21	8	0	23	0	-	-
...an	55,251	2,321	4.2%	9	7	1	0	3	2	1,544	5.3%
...d	198,115	27,538	13.9%	37	37	19	0	35	1	56,996	-7.1%
....	302,541	4,841	1.6%	17	11	12	3	29	13	39,479	2.2%
...nistan	188,457	7,915	4.2%	13	6	2	0	8	5	14,498	-
...Arab Emirates	32,278	-	-	4	8	1	0	6	0	1,239	32.1%
...stan	172,742	3,455	2.0%	9	9	2	0	4	1	7,602	2.4%
...n	128,066	4,738	3.7%	42	37	24	1	22	0	37,911	5.5%
....	203,850	-	-	6	12	2	0	10	2	1,734	-17.0%
....	919,595	45,980	5.0%	13	6	2	0	9	12	8,282	14.2%
....	481,354	31,769	6.6%	19	15	4	0	8	6	269,329	-1.7%
...na	43,484	4,957	11.4%	9	2	1	0	7	0	10,232	-20.9%
...na	224,607	41,552	18.5%	7	7	0	0	0	0	47,981	-8.7%
...a Faso	105,869	12,175	11.5%	7	2	1	0	0	0	27,371	-2.1%
...i	10,745	612	5.7%	6	7	0	0	0	3	363	-61.0%
...oon	183,568	8,261	4.5%	38	15	1	1	34	4	92,116	-8.5%
...erde	1,557	-	-	3	2	0	0	13	0	328	142.9%
...African Republic	240,536	20,927	8.7%	14	3	1	0	0	0	88,444	-1.3%
....	495,755	45,114	9.1%	15	5	1	0	0	1	49,004	-6.0%
...os	863	-	-	2	9	2	0	3	4	31	-33.3%
...Democratic Republic of the	132,047	6,602	5.0%	15	3	1	0	9	1	85,174	-0.8%
...lvoire	905,446	58,854	6.5%	40	28	2	0	9	45	522,037	-3.8%
...ti	124,504	7,470	6.0%	19	12	2	1	10	1	27,479	-27.1%
....	8,958	-	-	5	5	0	0	9	0	23	-
....	386,662	37,506	9.7%	13	7	6	0	13	1	278	38.5%
...rial Guinea	10,831	-	-	16	5	2	1	7	2	6,765	-5.7%
....	45,406	1,952	4.3%	12	7	6	0	8	0	6,120	-3.3%
...a	426,373	72,057	16.9%	35	16	1	0	0	4	17,734	-8.1%
....	103,347	723	0.7%	14	5	1	0	11	1	84,271	-0.5%
...a, The	4,127	95	2.3%	3	2	1	0	10	0	1,857	10.3%
....	92,098	5,157	5.6%	14	8	2	0	7	0	24,460	-15.9%
....	94,926	664	0.7%	12	10	1	1	7	3	26,753	-4.8%
...-Bissau	13,948	-	-	3	10	1	0	9	1	8,444	-9.0%
....	224,961	17,997	8.0%	50	24	5	0	27	15	66,008	-5.2%
...o	11,720	23	0.2%	6	7	0	0	1	1	54	-
....	43,000	731	1.7%	16	11	2	0	7	2	13,440	-17.9%
....	679,362	679	0.1%	8	1	3	0	8	0	1,382	15.1%
...ascar	226,658	9,746	4.3%	50	27	18	2	25	32	45,278	-9.1%
...i	45,747	5,124	11.2%	8	11	0	0	0	8	9,892	-21.6%
...ania	478,841	17,717	3.7%	13	4	1	0	1	0	50,911	-7.0%
...ania	397,956	6,765	1.7%	10	2	2	0	10	1	1,224	-23.6%
...ius	788	-	-	3	9	4	0	7	32	62	-5.9%
...co	172,414	1,207	0.7%	16	9	2	0	10	8	11,680	-0.4%
...bique	309,496	25,998	8.4%	15	16	5	0	19	7	118,151	-2.0%
...a	317,818	43,223	13.6%	14	11	3	1	11	1	31,043	-8.4%
....	489,192	37,668	7.7%	11	3	0	0	0	1	5,127	-31.7%
....	356,669	11,770	3.3%	27	9	2	0	11	1	52,189	-22.8%
...a	10,169	630	6.2%	8	9	0	0	0	2	1,185	-32.8%
...me and Principe	372	-	-	3	9	1	0	6	2	104	-
...al	75,951	8,810	11.6%	12	4	6	0	17	0	23,958	-6.8%
...eone	27,699	582	2.1%	12	10	3	0	7	4	4,073	-25.5%
...a	246,201	1,970	0.8%	19	10	2	0	16	1	29,016	-9.3%
...Africa	470,693	25,888	5.5%	36	28	19	9	47	113	34,429	-0.9%
....	967,500	50,310	5.2%	22	6	2	0	7	1	237,943	-13.5%
...and	6,704	-	-	5	5	0	0	0	0	2,015	12.5%
...ia	364,900	108,740	29.8%	41	33	5	0	26	47	149,850	-2.3%
....	21,925	1,732	7.9%	9	0	2	0	7	0	1,969	-29.1%
....	63,170	190	0.3%	11	5	3	0	8	5	1,969	2.2%
...a	93,065	22,894	24.6%	20	13	0	0	27	10	16,178	-17.9%
...a	290,586	92,697	31.9%	11	11	0	0	0	6	120,641	-21.4%
...owe	150,873	18,256	12.1%	11	10	0	0	0	2	73,514	-14.4%
...IA											
...ia	2,969,910	397,968	13.4%	63	35	38	35	74	282	596,678	-1.8%
....	7,056	78	1.1%	5	13	6	1	8	2	3,147	-2.0%
....	313	-	-	0	4	1	0	4	1	108	-
...esia, Federated States of ...	271	-	-	6	5	2	0	6	4	58	-37.5%
...ealand	104,454	30,918	29.6%	8	63	11	1	16	13	30,680	5.2%
...New Guinea	178,704	4,110	2.3%	58	32	9	0	31	12	118,151	-3.6%
....	1,093	-	-	3	8	1	0	4	1	405	-19.2%
...on Islands	10,954	33	0.3%	20	23	4	0	4	6	9,792	-1.7%
....	251	-	-	2	3	2	0	3	2	15	-
...u	4,707	-	-	5	8	2	0	4	0	1,726	1.4%

...ble presents data for most independent nations having an area greater than 200 square miles
...insignificant, or not available
...rce: World Resources Institute, 2003. Earth Trends: The Environmental Information Portal. Available at http://earthtrends.wri.org. Washington D. C. World Resources Institute
...rce: United Nations Environment Programme - World Conservation Monitoring Centre (UNEP-WCMC); World Database on Protected Areas
...rce: International Union of Conservation of Nature and Natural Resources; IUCN 2003 Red List of Threatened Species <www.redlist.org>
...rce: United Nations Food and Agriculture Organization; Global Forest Resources Assessment 2000

WORLD COMPARISONS

General Information

Equatorial diameter of the earth, 7,926.38 miles.
Polar diameter of the earth, 7,899.80 miles.
Mean diameter of the earth, 7,917.52 miles.
Equatorial circumference of the earth, 24,901.46 miles.
Polar circumference of the earth, 24,855.34 miles.
Mean distance from the earth to the sun, 93,020,000 miles.
Mean distance from the earth to the moon, 238,857 miles.
Total area of the earth, 197,000,000 sq. miles.

Highest elevation on the earth's surface, Mt. Everest, Asia, 29,028 ft.
Lowest elevation on the earth's land surface, shores of the Dead Sea, Asia, 1,339 ft. below sea level.
Greatest known depth of the ocean, southwest of Guam, Pacific Ocean, 35,810 ft.
Total land area of the earth (incl. inland water and Antarctica), 57,900,000 sq. miles.

Area of Africa, 11,700,000 sq. miles.
Area of Antarctica, 5,400,000 sq. miles.
Area of Asia, 17,300,000 sq. miles.
Area of Europe, 3,800,000 sq. miles.
Area of North America, 9,500,000 sq. miles.
Area of Oceania (incl. Australia) 3,300,000 sq. miles
Area of South America, 6,900,000 sq. miles.
Population of the earth (est. 1/1/04), 6,339,505,000

Principal Islands and Their Areas

ISLAND — **Area (Sq. Mi.)**

Baffin I., Canada ... 195,928
Banks I., Canada ... 27,038
Borneo (Kalimantan), Asia ... 287,300
Bougainville, Papua New Guinea ... 3,591
Cape Breton I., Canada ... 3,981
Celebes (Sulawesi), Indonesia ... 73,057
Ceram (Seram), Indonesia ... 7,191
Corsica, France ... 3,367
Crete, Greece ... 3,189
Cuba, N. America ... 42,780
Cyprus, Asia ... 3,572
Devon I., Canada ... 21,331
Ellesmere I., Canada ... 75,767

Flores, Indonesia ... 5,502
Great Britain, U.K. ... 88,795
Greenland, N. America ... 840,000
Guadalcanal, Solomon Is. ... 2,060
Hainan Dao, China ... 13,127
Hawaii, U.S. ... 4,028
Hispaniola, N. America ... 29,300
Hokkaidō, Japan ... 32,245
Honshū, Japan ... 89,176
Iceland, Europe ... 39,769
Ireland, Europe ... 32,587
Jamaica, N. America ... 4,247
Java (Jawa), Indonesia ... 51,038
Kodiak I., U.S. ... 3,670

Kyūshū, Japan ... 17,129
Lyete, Philippines ... 2,785
Long Island, U.S. ... 1,377
Luzon, Philippines ... 40,420
Madagascar, Africa ... 226,642
Melville I., Canada ... 16,274
Mindanao, Philippines ... 36,537
Mindoro, Philippines ... 3,759
Negros, Philippines ... 4,907
New Britain, Papua New Guinea ... 14,093
New Caledonia, Oceania ... 6,252
Newfoundland, Canada ... 42,031
New Guinea, Asia-Oceania ... 308,882

New Ireland, Papua New Guinea ... 3,475
North East Land, Norway ... 6,350
North I., New Zealand ... 44,333
Novaya Zemlya, Russia ... 31,892
Palawan, Philippines ... 4,550
Panay, Philippines ... 4,446
Prince of Wales I., Canada ... 12,872
Puerto Rico, N. America ... 3,514
Sakhalin, Russia ... 29,498
Samar, Philippines ... 5,050
Sardinia, Italy ... 9,301
Shikoku, Japan ... 7,258
Sicily, Italy ... 9,926

Somerset I., Canada ...
Southampton I., Canada ...
South I., New Zealand ...
Spitsbergen, Norway ...
Sri Lanka, Asia ...
Sumatra (Sumatera), Indonesia ...
Taiwan, Asia ...
Tasmania, Australia ...
Tierra del Fuego, S. America ...
Timor, Asia ...
Vancouver I., Canada ...
Victoria I., Canada ...
Vrangelya (Wrangel), Russia ...

Principal Lakes, Oceans, Seas, and Their Areas

LAKE Country — **Area (Sq. Mi.)**

Arabian Sea ... 1,492,000
Aral Sea, Kazakhstan-Uzbekistan ... 13,000
Arctic Ocean ... 5,400,000
Athabasca, L., Canada ... 3,064
Atlantic Ocean ... 29,600,000
Balqash köli (L. Balkhash), Kazakhstan ... 7,027
Baltic Sea, Europe ... 163,000
Baykal, Ozero (L. Baikal), Russia ... 12,162
Bering Sea, Asia-N.A. ... 876,000

Black Sea, Europe-Asia ... 178,000
Caribbean Sea, N.A.-S.A. ... 1,063,000
Caspian Sea, Asia-Europe ... 144,402
Chad, L., Cameroon-Chad-Nigeria ... 595
Erie, L., Canada-U.S. ... 9,910
Eyre, L., Australia ... 3,668
Gairdner, L., Australia ... 1,076
Great Bear Lake, Canada ... 12,096
Great Salt Lake, U.S. ... 1,700
Great Slave Lake, Canada ... 11,030

Hudson Bay, Canada ... 475,000
Huron, L., Canada-U.S. ... 23,000
Indian Ocean ... 26,500,000
Japan, Sea of, Asia ... 389,000
Koko Nor (Qinghai Hu), China ... 1,722
Ladozhskoye Ozero (L. Ladoga), Russia ... 7,002
Manitoba, L., Canada ... 1,785
Mediterranean Sea, Europe-Africa-Asia ... 967,000
Mexico, Gulf of, N. America ... 596,000

Michigan, L., U.S. ... 22,300
Nicaragua, Lago de, Nicaragua ... 3,147
North Sea, Europe ... 222,000
Nyasa, L., Malawi-Mozambique-Tanzania ... 11,120
Onezhskoye Ozero (L. Onega), Russia ... 3,819
Ontario, L., Canada-U.S. ... 7,340
Pacific Ocean ... 60,100,000
Red Sea, Africa-Asia ... 169,000
Rudolf, L., Ethiopia-Kenya ... 2,471

Southern Ocean ... 7,8
Superior, L., Canada-U.S. ...
Tanganyika. L., Africa ...
Titicaca, Lago, Bolivia-Peru ...
Torrens, L., Australia ...
Vänern (L.), Sweden ...
Van Gölü (L.), Turkey ...
Victoria, L., Kenya-Tanzania-Uganda ...
Winnipeg, L., Canada ...
Winnipegosis, L., Canada ...
Yellow Sea, China-Korea ... 4

Principal Mountains and Their Heights

MOUNTAIN Country — **Elev. (Ft.)**

Aconcagua, Cerro, Argentina ... 22,831
Annapurna, Nepal ... 26,504
Aoraki, New Zealand ... 12,316
Api, Nepal ... 23,399
Apo, Philippines ... 9,692
Ararat, Mt., Turkey ... 16,854
Barú, Volcán, Panama ... 11,401
Bangueta, Mt., Papua New Guinea ... 13,520
Belukha, Mt., Kazakhstan-Russia ... 14,783
Bia, Phou, Laos ... 9,249
Blanc, Mont (Monte Bianco), France-Italy ... 15,771
Blanca Pk., Colorado, U.S. ... 14,345
Bolívar, Pico, Venezuela ... 16,427
Bonete, Cerro, Argentina ... 22,546
Borah Pk., Idaho, U.S. ... 12,662
Boundary Pk., Nevada, U.S. ... 13,140
Cameroon Mtn., Cameroon ... 13,451
Carrauntoohil, Ireland ... 3,406
Chaltel, Cerro (Monte Fitzroy), Argentina-Chile ... 10,958
Chimborazo, Ecuador ... 20,702
Chirripó, Cerro, Costa Rica ... 12,530
Colima, Nevado de, Mexico ... 13,911
Cotopaxi, Ecuador ... 19,347
Cristóbal Colón, Pico, Colombia ... 19,029
Damāvand, Qolleh-ye, Iran ... 18,386
Dhawalāgiri, Nepal ... 26,810
Duarte, Pico, Dominican Rep. ... 10,417
Dufourspitze (Monte Rosa), Italy-Switzerland ... 15,203
Elbert, Mt., Colorado, U.S. ... 14,433
El'brus, Gora, Russia ... 18,510

Elgon, Mt., Kenya-Uganda ... 14,178
Erciyeş, Dağı, Turkey ... 12,848
Etna, Mt., Italy ... 10,902
Everest, Mt., China-Nepal ... 29,028
Fairweather, Mt., Alaska-Canada ... 15,300
Folādī, Koh-e, Afghanistan ... 16,847
Foraker, Mt., Alaska, U.S. ... 17,400
Fuji San, Japan ... 12,388
Galdhøpiggen, Norway ... 8,100
Gannett Pk., Wyoming, U.S. ... 13,804
Gasherbrum, China-Pakistan ... 26,470
Gerlachovský štít, Slovakia ... 8,711
Giluwe, Mt., Papua New Guinea ... 14,331
Gongga Shan, China ... 24,790
Grand Teton, Wyoming, U.S. ... 13,770
Grossglockner, Austria ... 12,457
Hadūr Shu'ayb, Yemen ... 12,008
Haleakalā Crater, Hawaii, U.S. ... 10,023
Hekla, Iceland ... 4,892
Hood, Mt., Oregon, U.S. ... 11,239
Huascarán, Nevado, Peru ... 22,133
Huila, Nevado de, Colombia ... 18,865
Hvannadalshnúkur, Iceland ... 6,952
Illampu, Nevado, Bolivia ... 21,066
Illimani, Nevado, Bolivia ... 20,741
Ismail Samani, pik, Tajikistan ... 24,590
Iztaccíhuatl, Mexico ... 17,159
Jaya, Puncak, Indonesia ... 16,503
Jungfrau, Switzerland ... 13,642
K2 (Qogir Feng), China-Pakistan ... 28,250
Kāmet, China-India ... 25,447
Kānchenjunga, India-Nepal ... 28,208
Kātrīnā, Jabal, Egypt ... 8,668

Kebnekaise, Sweden ... 6,926
Kenya, Mt. (Kirinyaga), Kenya ... 17,058
Kerinci, Gunung, Indonesia ... 12,467
Kilimanjaro, Tanzania ... 19,340
Kinabalu, Gunong, Malaysia ... 13,455
Klyuchevskaya, Russia ... 15,584
Kosciuszko, Mt., Australia ... 7,313
Koussi, Emi, Chad ... 11,204
Kula Kangri, Bhutan ... 24,784
La Selle, Massif de, Haiti ... 8,793
Lassen Pk., California, U.S. ... 10,457
Llullaillaco, Volcán, Argentina-Chile ... 22,110
Logan, Mt., Canada ... 19,551
Longs Pk., Colorado, U.S. ... 14,255
Makālu, China-Nepal ... 27,825
Margherita Peak, Dem. Rep. of the Congo-Uganda ... 16,763
Markham, Mt., Antarctica ... 14,049
Maromokotro, Madagascar ... 9,436
Massive, Mt., Colorado, U.S. ... 14,421
Matterhorn, Italy-Switzerland ... 14,692
Mauna Kea, Hawaii, U.S. ... 13,796
Mauna Loa, Hawaii, U.S. ... 13,679
Mayon Volcano, Philippines ... 8,077
McKinley, Mt., Alaska, U.S. ... 20,320
Meron, Hare, Israel ... 3,963
Meru, Mt., Tanzania ... 14,978
Misti, Volcán, Peru ... 19,101
Mitchell, Mt., North Carolina, U.S. ... 6,684
Môco, Serra do, Angola ... 8,596
Moldoveanu, Romania ... 8,346
Mulhacén, Spain ... 11,424

Musala, Bulgaria ... 9,596
Muztag, China ... 25,338
Muztagata, China ... 24,757
Namjagbarwa Feng, China ... 25,446
Nanda Devi, India ... 25,645
Nanga Parbat, Pakistan ... 26,660
Narodnaya, Gora, Russia ... 6,217
Nevis, Ben, United Kingdom ... 4,406
Ojos del Salado, Nevado, Argentina-Chile ... 22,615
Ólimbos, Cyprus ... 6,401
Ólympos, Greece ... 9,570
Olympus, Mt., Washington, U.S. ... 7,965
Orizaba, Pico de, Mexico ... 18,406
Paektu San, North Korea-China ... 9,003
Paricutín, Mexico ... 9,186
Parnassós, Greece ... 8,061
Pelée, Montagne, Martinique ... 4,583
Pidurutalagala, Sri Lanka ... 8,281
Pikes Pk., Colorado, U.S. ... 14,110
Pobedy, pik, China-Kyrgyzstan ... 24,406
Popocatépetl, Volcán, Mexico ... 17,930
Pulog, Mt., Philippines ... 9,626
Rainier, Mt., Washington, U.S. ... 14,410
Ramm, Jabal, Jordan ... 5,755
Ras Dashen Terara, Ethiopia ... 15,158
Rinjani, Gunung, Indonesia ... 12,224
Robson, Mt., Canada ... 12,972
Roraima, Mt., Brazil-Guyana-Venezuela ... 9,432
Ruapehu, Mt., New Zealand ... 9,177

St. Elias, Mt., Alaska, U.S.-Canada ...
Sajama, Nevado, Bolivia ...
Semeru, Gunung, Indonesia ...
Shām, Jabal ash, Oman ...
Shasta, Mt., California, U.S. ...
Snowdon, United Kingdom ...
Tahat, Algeria ...
Tajumulco, Guatemala ...
Taranaki, Mt., New Zealand ...
Tirich Mīr, Pakistan ...
Tomanivi (Victoria), Fiji ...
Toubkal, Jebel, Morocco ...
Triglav, Slovenia ...
Trikora, Puncak, Indonesia ...
Tupungato, Cerro, Argentina-Chile ...
Turquino, Pico, Cuba ...
Uluru (Ayers Rock), Australia ...
Uncompahgre Pk, Colorado, U.S. ...
Vesuvio (Vesuvius), Italy ...
Victoria, Mt., Papua New Guinea ...
Vinson Massif, Antarctica ...
Waddington, Mt., Canada ...
Washington, Mt., New Hampshire, U.S. ...
Whitney, Mt., California, U.S. ...
Wilhelm, Mt., Papua New Guinea ...
Wrangell, Mt., Alaska, U.S. ...
Xixabangma Feng (Gosainthan), China ...
Yü Shan, Taiwan ...
Zugspitze, Austria-Germany ...

Principal Rivers and Their Lengths

RIVER Continent — **Length (Mi.)**

Albany, N. America ... 610
Aldan, Asia ... 1,412
Amazonas-Ucayali, S. America ... 4,000
Amu Darya, Asia ... 1,578
Amur, Asia ... 1,752
Araguaia, S. America ... 1,367
Arkansas, N. America ... 1,460
Atchafalaya, N. America ... 1,420
Athabasca, N. America ... 765
Brahmaputra, Asia ... 1,770
Brazos, N. America ... 1,280
Canadian, N. America ... 906
Churchill, N. America ... 1,000
Colorado, N. America (U.S.-Mexico) ... 1,450
Colorado, N. America (Texas) ... 862
Columbia, N. America ... 1,240
Congo (Zaïre), Africa ... 2,715
Danube, Europe ... 1,777
Darling, Australia ... 864
Dnieper (Dnipro), Europe ... 1,367

Don, Europe ... 1,162
Elbe, Europe ... 690
Essequibo, S. America ... 603
Euphrates, Asia ... 1,510
Fraser, N. America ... 851
Ganges, Asia ... 1,864
Gila, N. America ... 649
Godāvari, Asia ... 932
Huang (Yellow), Asia ... 2,902
Indigirka, Asia ... 1,072
Indus, Asia ... 1,118
Irrawaddy, Asia ... 1,300
Juruá, S. America ... 1,250
Kama, Europe ... 1,122
Kasai, Africa ... 1,338
Kolyma, Asia ... 1,323
Lena, Asia ... 2,734
Limpopo, Africa ... 1,100
Loire, Europe ... 690
Mackenzie, N. America ... 2,635
Madeira, S. America ... 2,013
Magdalena, S. America ... 951

Marañón, S. America ... 1,000
Mekong, Asia ... 2,796
Meuse, Europe ... 575
Mississippi, N. America ... 2,340
Mississippi-Missouri, N. America ... 3,710
Missouri, N. America ... 2,540
Murray-Darling, Australia ... 2,169
Negro, S. America ... 1,305
Nelson, N. America ... 1,600
Niger, Africa ... 2,585
Nile, Africa ... 4,132
Ob', Asia ... 2,268
Oder, Europe ... 565
Ohio, N. America ... 1,310
Oka, Europe ... 932
Orange, Africa ... 1,300
Orinoco, S. America ... 1,703
Ottawa, N. America ... 790
Paraguay, S. America ... 1,610
Parnaíba, S. America ... 901
Peace, N. America ... 1,195

Pechora, Europe ... 1,125
Pecos, N. America ... 926
Pilcomayo, S. America ... 1,550
Plata-Paraná, S. America ... 2,920
Platte, N. America ... 990
Purús, S. America ... 1,860
Red, N. America ... 1,290
Rhine, Europe ... 820
Rhône, Europe ... 503
Rio Grande, N. America ... 1,900
Roosevelt, S. America ... 950
St. Lawrence, N. America ... 1,900
Salado, S. America ... 870
Salween (Nu), Asia ... 1,750
São Francisco, S. America ... 1,740
Saskatchewan-Bow, N. America ... 1,205
Severnaya Dvina (Northern Dvina), Europe ... 462
Snake, N. America ... 1,040
Sungari (Songhua), Asia ... 1,140
Syr Darya, Asia ... 1,370

Tagus, Europe ...
Tarim, Asia ...
Tennessee, N. America ...
Tigris, Asia ...
Tisa, Europe ...
Tocantins, S. America ...
Ucayali, S. America ...
Ural, Asia ...
Uruguay, S. America ...
Verkhnyaya Tunguska (Angara), Asia ...
Vilyuy, Asia ...
Volga, Europe ...
Volta, Africa ...
Wisła (Vistula), Europe ...
Xiang, Asia ...
Xingú, S. America ...
Yangtze (Chang), Asia ...
Yellowstone, N. America ...
Yenisey, Asia ...
Yukon, N. America ...
Zambezi, Africa ...

INCIPAL CITIES OF THE WORLD

n, Cote d'Ivoire ...1,929,079
ıby (Abu Dhabi), United Arab
tes ...242,975
Ghana (1,390,000) ...949,113
Ababa, Ethiopia ...2,424,000
dābād, India (4,519,278) ...3,515,361
ɔ (Ḩalab), Syria (1,640,000) ...1,591,400
ndria (Al Iskandarīyah), Egypt
),000) ...3,339,076
s (El Djazaïr), Algeria
7,983) ...1,507,241
h (Giza), Egypt
Qāhirah) ...2,221,817
y, Kazakhstan (1,190,000) ...1,129,356
ān, Jordan (1,500,000) ...1,147,447
rdam, Netherlands
1,303) ...727,053
a, Turkey (3,294,220) ...2,984,099
anarivo, Madagascar ...1,250,000
rp (Antwerpen), Belgium
5,000) ...453,030
bat (Ashkhabad),
nenistan ...557,600
a, Eritrea ...358,100
a (Aqmola), Kazakhstan
324) ...312,965
ión, Paraguay (700,000) ...546,637
s (Athína), Greece (3,150,000) ...772,072
a, Georgia, U.S. (4,112,198) ...416,474
and, New Zealand (1,074,510) ...367,737
ād, Iraq ...3,841,268
'Bakı), Azerbaijan
),000) ...1,792,300
ko, Mali ...658,275
ng, Indonesia ...5,919,400
lore, India (5,686,844) ...4,292,223
āzī, Libya ...800,000
ok (Krung Thep), Thailand
),000) ...5,620,591
i, Central African Republic ...451,690
ona, Spain (4,000,000) ...1,496,266
, China (7,320,000) ...6,690,000
(Bayrūt), Lebanon (1,675,000) ...509,000
t, N. Ireland, U.K. (730,000) ...297,300
de (Beograd), Serbia and
tenegro ...1,594,483
orizonte, Brazil (4,055,000) ...1,366,301
Germany (4,220,000) ...3,386,667
gham, England, U.K.
5,000) ...1,020,589
k, Kyrgyzstan ...753,400
á, Colombia ...6,422,198
Germany (600,000) ...301,048
n, Massachusetts, U.S.
9,100) ...589,141
a, Brazil ...1,947,133
ava, Slovakia ...451,395
ville, Congo ...693,712
ne, Australia (1,627,535) ...888,449
ls (Bruxelles), Belgium
),000) ...133,845
est (București), Romania
),000) ...2,016,131
est, Hungary (2,450,000) ...1,825,153
s Aires, Argentina
00,000) ...2,960,976
(Al Qāhirah), Egypt
),000) ...6,800,992
y, Alberta, Canada (951,395) ...878,866
olombia ...2,128,920
rra, Australia (342,798) ...311,518
Town, South Africa
),000) ...854,616
s, Venezuela (4,000,000) ...1,822,465
f, Wales, U.K. (645,000) ...315,040
anca, Morocco (3,400,000) ...3,022,000
chun, China ...2,470,000
abinsk, Russia (1,320,000) ...1,086,300
du, China ...2,760,000
ai (Madras), India
4,624) ...4,216,268
jo, Illinois, U.S. (9,157,540) ...2,896,016
ău (Kishinev), Moldova
500) ...658,300
gong, Bangladesh
2,662) ...1,566,070
qing, China ...3,870,000
nati, Ohio, U.S. (1,979,202) ...331,285
and, Ohio, U.S. (2,945,831) ...478,403
ne (Köln), Germany
),000) ...962,507
bo, Sri Lanka (2,050,000) ...615,000
ry, Guinea ...950,000
hagen (København), Denmark
),000) ...499,148
ba, Argentina (1,260,000) ...1,179,067

Cotonou, Benin ...650,660
Curitiba, Brazil (2,595,000) ...1,586,848
Dakar, Senegal (1,976,533) ...879,703
Dalian, China ...2,400,000
Dallas, Texas, U.S. (5,221,801) ...1,188,580
Damascus (Dimashq), Syria
(2,230,000) ...1,549,932
Dar es Salaam, Tanzania ...1,360,850
Delhi, India (12,791,458) ...9,817,439
Denver, Colorado, U.S. (2,581,506) ...554,636
Detroit, Michigan, U.S. (5,456,428) ...951,270
Dhaka (Dacca), Bangladesh
(6,537,308) ...3,637,892
Djibouti, Djibouti ...329,337
Dnipropetrovs'k, Ukraine
(1,590,000) ...1,108,682
Donets'k, Ukraine (2,090,000) ...1,050,369
Douala, Cameroon ...712,251
Dublin (Baile Átha Cliath), Ireland
(1,175,000) ...481,854
Durban, South Africa (1,740,000) ...669,242
Dushanbe, Tajikistan (700,000) ...528,600
Düsseldorf, Germany (1,200,000) ...568,855
Edinburgh, Scotland, U.K. (640,000) ...448,850
Edmonton, Alberta, Canada
(937,845) ...666,104
Eşfahān, Iran (1,525,000) ...1,266,072
Essen, Germany (5,040,000) ...599,515
Fortaleza, Brazil (2,780,000) ...788,956
Frankfurt am Main, Germany
(1,960,000) ...643,821
Fukuoka, Japan (2,000,000) ...1,341,489
Geneva (Génève), Switzerland
(450,592) ...172,598
Glasgow, Scotland, U.K. (1,870,000) ...616,430
Goiânia, Brazil ...1,075,761
Guadalajara, Mexico (3,669,021) ...1,646,183
Guangzhou (Canton), China ...3,750,000
Guatemala, Guatemala
(1,500,000) ...1,006,954
Guayaquil, Ecuador ...2,117,553
Halifax, Nova Scotia, Canada
(359,183) ...119,300
Hamburg, Germany (2,460,000) ...1,704,735
Hannover, Germany (1,015,000) ...514,718
Hanoi, Vietnam (1,275,000) ...1,073,760
Harare, Zimbabwe (1,470,000) ...1,189,103
Harbin, China ...3,120,000
Havana (La Habana), Cuba
(2,285,000) ...2,189,716
Helsinki, Finland (939,697) ...548,720
Hiroshima, Japan (1,600,000) ...1,126,282
Ho Chi Minh City (Saigon), Vietnam
(3,300,000) ...3,015,743
Hong Kong (Xianggang), China
(4,770,000) ...1,250,993
Honolulu, Hawaii, U.S. (876,156) ...371,657
Houston, Texas, U.S. (4,669,571) ...1,953,631
Hyderābād, India (5,533,640) ...3,449,878
Ibadan, Nigeria ...1,144,000
Islāmābād, Pakistan (*Rāwalpindi) ...529,180
İstanbul, Turkey (8,506,026) ...8,260,438
İzmir, Turkey (2,554,363) ...2,081,556
Jaipur, India ...2,324,319
Jakarta, Indonesia (10,200,000) ...9,373,900
Jerusalem (Yerushalayim), Israel
(685,000) ...633,700
Jiddah, Saudi Arabia ...1,450,000
Jinan, China ...2,150,000
Johannesburg, South Africa
(4,000,000) ...752,349
Kābul, Afghanistan ...1,424,400
Kampala, Uganda ...773,463
Kānpur, India (2,690,486) ...2,540,069
Kaohsiung, Taiwan (1,845,000) ...1,468,586
Karāchi, Pakistan ...9,339,023
Katowice, Poland (2,755,000) ...343,158
Kharkiv, Ukraine (1,950,000) ...1,494,235
Khartoum (Al Kharţūm), Sudan
(1,450,000) ...947,483
Kiev (Kyïv), Ukraine (3,250,000) ...2,589,541
Kingston, Jamaica (830,000) ...516,500
Kinshasa, Dem. Rep. of
the Congo ...3,000,000
Kitakyūshū, Japan (1,550,000) ...1,011,491
Kolkata (Calcutta), India
(13,216,546) ...4,580,544
Kuala Lumpur, Malaysia
(2,500,000) ...1,297,526
Kuwait (Al Kuwayt), Kuwait
(1,126,000) ...28,859
Lagos, Nigeria (3,800,000) ...1,213,000
Lahore, Pakistan ...5,143,495
La Paz, Bolivia (1,487,854) ...792,611
Libreville, Gabon (418,616) ...362,386
Lilongwe, Malawi ...435,964

Lima, Peru (6,321,173) ...340,422
Lisbon (Lisboa), Portugal (2,350,000) ...563,210
Liverpool, England, U.K. (1,515,000) ...467,995
Ljubljana, Slovenia ...263,832
Lomé, Togo ...450,000
London, England, U.K.
(12,000,000) ...7,074,265
Los Angeles, California, U.S.
(16,373,645) ...3,694,820
Luanda, Angola ...1,459,900
Lucknow, India (2,266,933) ...2,207,340
Lusaka, Zambia ...1,269,848
Lyon, France (1,648,216) ...445,452
Madrid, Spain (4,690,000) ...2,882,860
Managua, Nicaragua ...864,201
Manaus, Brazil ...1,394,724
Manchester, England, U.K.
(2,760,000) ...430,818
Manila, Philippines (11,200,000) ...1,654,761
Mannheim, Germany (1,525,000) ...307,730
Maputo, Mozambique ...966,837
Maracaibo, Venezuela ...1,249,670
Marseille, France (1,516,340) ...798,430
Mashhad, Iran ...1,887,405
Mecca (Makkah), Saudi Arabia ...630,000
Medan, Indonesia ...1,988,200
Medellín, Colombia (2,290,000) ...1,885,001
Melbourne, Australia (3,366,542) ...67,784
Mexico City (Ciudad de México),
Mexico (17,786,983) ...8,605,239
Miami, Florida, U.S. (3,876,380) ...362,470
Milan (Milano), Italy (3,790,000) ...1,305,591
Milwaukee, Wisconsin, U.S.
(1,689,572) ...596,974
Minneapolis, Minnesota, U.S.
(2,968,806) ...382,618
Minsk, Belarus (1,680,567) ...1,677,137
Mogadishu (Muqdisho), Somalia ...600,000
Monrovia, Liberia ...465,000
Monterrey, Mexico (3,236,604) ...1,110,909
Montevideo, Uruguay (1,650,000) ...1,303,182
Montréal, Quebec, Canada
(3,426,350) ...1,039,534
Moscow (Moskva), Russia
(12,850,000) ...8,389,700
Mumbai (Bombay), India
(16,368,084) ...11,914,398
Munich (München), Germany
(1,930,000) ...1,194,560
Nagoya, Japan (5,250,000) ...2,171,378
Nāgpur, India (2,122,965) ...2,051,320
Nairobi, Kenya ...2,143,254
Nanjing, China ...2,490,000
Naples (Napoli), Italy (3,150,000) ...1,046,987
N'Djamena, Chad ...546,572
Newcastle upon Tyne, England, U.K.
(1,350,000) ...282,338
New Delhi, India (*Delhi) ...294,783
New York, New York, U.S.
(21,199,865) ...8,008,278
Niamey, Niger ...392,165
Nizhniy Novgorod, Russia
(1,950,000) ...1,364,900
Nouakchott, Mauritania ...393,325
Novosibirsk, Russia (1,505,000) ...1,402,400
Nürnberg, Germany (1,065,000) ...486,628
Odesa, Ukraine (1,150,000) ...1,002,246
Omsk, Russia (1,190,000) ...1,157,600
Ōsaka, Japan (17,050,000) ...2,598,589
Oslo, Norway (773,498) ...504,040
Ottawa, Ontario, Canada
(1,063,664) ...774,072
Ouagadougou, Burkina Faso ...634,479
Palembang, Indonesia ...1,415,500
Panamá, Panama (995,000) ...415,964
Paris, France (11,174,743) ...2,125,246
Patna, India (1,707,429) ...1,376,950
Perm', Russia (1,110,000) ...1,017,100
Perth, Australia (1,244,320) ...10,195
Philadelphia, Pennsylvania, U.S.
(6,188,463) ...1,517,550
Phnom Penh (Phnum Pénh),
Cambodia ...570,155
Phoenix, Arizona, U.S. (3,251,876) ...1,321,045
Port Moresby, Papua New Guinea ...246,664
Port-au-Prince, Haiti (1,425,594) ...990,558
Portland, Oregon, U.S. (2,265,223) ...529,121
Porto, Portugal (1,230,000) ...273,060
Porto Alegre, Brazil (3,375,000) ...1,304,998
Prague (Praha), Czech Republic
(1,328,000) ...1,193,270
Pretoria, South Africa (1,100,000) ...692,348
Pune, India (3,755,525) ...2,540,069
Pusan, South Korea ...3,814,325
P'yŏngyang, North Korea ...2,741,260
Qingdao, China ...2,300,000

Québec, Quebec, Canada (682,757) ...169,076
Quezon City, Philippines
(*Manila) ...1,989,419
Quito, Ecuador ...1,615,809
Rabat, Morocco (1,200,000) ...717,000
Rangoon (Yangon), Myanmar
(2,800,000) ...2,705,039
Recife, Brazil (3,160,000) ...1,421,993
Regina, Saskatchewan, Canada
(192,800) ...178,225
Reykjavík, Iceland (166,015) ...107,684
Rīga, Latvia (1,000,000) ...792,508
Rio de Janeiro, Brazil (10,465,000) ...5,851,914
Riyadh (Ar Riyāḑ), Saudi Arabia ...1,800,000
Rome (Roma), Italy (3,235,000) ...2,649,765
Rosario, Argentina (1,190,000) ...894,645
Rostov-na-Donu, Russia
(1,160,000) ...1,017,300
Rotterdam, Netherlands (1,089,979) ...539,000
Sacramento, California, U.S.
(1,796,857) ...407,018
St. Louis, Missouri, U.S. (2,603,607) ...348,189
St. Petersburg (Leningrad), Russia
(6,000,000) ...4,728,200
Salvador, Brazil (2,855,000) ...2,439,823
Samara, Russia (1,450,000) ...1,168,000
San Diego, California, U.S.
(2,813,833) ...1,223,400
San Francisco, California, U.S.
(7,039,362) ...776,733
San José, Costa Rica (996,194) ...309,672
San Juan, Puerto Rico (1,967,627) ...421,958
San Salvador, El Salvador
(1,908,921) ...473,372
Santiago, Chile ...4,788,543
Santo Domingo, Dominican
Republic ...2,677,056
São Paulo, Brazil (17,380,000) ...9,713,692
Sapporo, Japan (2,000,000) ...1,822,300
Sarajevo, Bosnia and Herzegovina ...367,703
Saratov, Russia (1,135,000) ...881,000
Seattle, Washington, U.S.
(3,554,760) ...563,374
Seoul (Sŏul), South Korea
(15,850,000) ...10,231,217
Shanghai, China (11,010,000) ...8,930,000
Shenyang (Mukden), China ...4,050,000
Singapore, Singapore (4,400,000) ...4,017,700
Skopje, Macedonia ...440,577
Sofia (Sofiya), Bulgaria (1,189,794) .1,138,629
Stockholm, Sweden (1,643,366) ...743,703
Stuttgart, Germany (2,020,000) ...582,443
Surabaya, Indonesia ...2,801,300
Sūrat, India (2,811,466) ...2,433,787
Sydney, Australia (3,741,290) ...11,115
T'aipei, Taiwan (6,200,000) ...2,640,322
Tallinn, Estonia ...403,981
Tashkent (Toshkent), Uzbekistan
(2,325,000) ...2,142,700
Tbilisi, Georgia (1,460,000) ...1,279,000
Tegucigalpa, Honduras ...576,661
Tehrān, Iran (8,800,000) ...6,758,845
Tel Aviv-Yafo, Israel (1,890,000) ...348,100
Tianjin (Tientsin), China ...5,000,000
Tiranë, Albania ...244,153
Tōkyō, Japan (30,300,000) ...8,130,408
Toronto, Ontario, Canada
(4,682,897) ...2,481,494
Tripoli (Ţarābulus), Libya ...1,500,000
Tunis, Tunisia (1,300,000) ...702,330
Turin (Torino), Italy (1,550,000) ...921,485
Ufa, Russia (1,110,000) ...1,088,900
Ulan Bator (Ulaanbaatar),
Mongolia ...672,882
Ürümqi, China ...1,130,000
València, Spain (1,340,000) ...739,014
Vancouver, British Columbia, Canada
(1,986,965) ...545,671
Viangchan (Vientiane), Laos ...464,000
Vienna (Wien), Austria (1,950,000) ...1,609,631
Vilnius, Lithuania ...578,334
Volgograd (Stalingrad), Russia
(1,358,000) ...1,000,000
Warsaw (Warszawa), Poland
(2,300,000) ...1,615,369
Washington, D.C., U.S. (7,608,070) ...572,059
Wellington, New Zealand (346,500) ...167,400
Winnipeg, Manitoba, Canada
(671,274) ...619,544
Wuhan, China ...3,870,000
Xi'an, China ...2,410,000
Yekaterinburg, Russia (1,530,000) ...1,272,900
Yerevan, Armenia (1,315,000) ...1,249,202
Yokohama, Japan (*Tōkyō) ...3,426,506
Zagreb, Croatia ...867,865
Zürich, Switzerland (932,681) ...337,553

olitan area populations are shown in parentheses.
located within the metropolitan area of another city; for example, Yokohama, Japan is located in the Tōkyō metropolitan area.

GLOSSARY OF FOREIGN GEOGRAPHICAL TERMS

Annam . . . Annamese
Arab . . . Arabic
Bantu . . . Bantu
Bur . . . Burmese
Camb . . . Cambodian
Celt . . . Celtic
Chn . . . Chinese
Czech . . . Czech
Dan . . . Danish
Du . . . Dutch
Fin . . . Finnish
Fr . . . French
Ger . . . German
Gr . . . Greek
Hung . . . Hungarian
Ice . . . Icelandic
India . . . India
Indian . . . American Indian
Indon . . . Indonesian
It . . . Italian
Jap . . . Japanese
Kor . . . Korean
Mal . . . Malayan
Mong . . . Mongolian
Nor . . . Norwegian
Per . . . Persian
Pol . . . Polish
Port . . . Portuguese
Rom . . . Romanian
Rus . . . Russian
Siam . . . Siamese
So. Slav . . . Southern Slavonic
Sp . . . Spanish
Swe . . . Swedish
Tib . . . Tibetan
Tur . . . Turkish
Yugo . . . Yugoslav

å, Nor., Swe . . . brook, river
aa, Dan., Nor . . . brook
aas, Dan., Nor . . . ridge
âb, Per . . . water, river
abad, India, Per . . . town, city
ada, Tur . . . island
adrar, Berber . . . mountain
air, Indon . . . stream
akrotírion, Gr . . . cape
älf, Swe . . . river
alp, Ger . . . mountain
altipiano, It . . . plateau
alto, Sp . . . height
archipel, Fr . . . archipelago
archipiélago, Sp . . . archipelago
arquipélago, Port . . . archipelago
arroyo, Sp . . . brook, stream
ås, Nor., Swe . . . ridge
austral, Sp . . . southern
baai, Du . . . bay
bab, Arab . . . gate, port
bach, Ger . . . brook, stream
backe, Swe . . . hill
bad, Ger . . . bath, spa
bahía, Sp . . . bay, gulf
bahr, Arab . . . river, sea, lake
baia, It . . . bay, gulf
baía, Port . . . bay
baie, Fr . . . bay, gulf
bajo, Sp . . . depression
bak, Indon . . . stream
bakke, Dan., Nor . . . hill
balkan, Tur . . . mountain range
bana, Jap . . . point, cape
banco, Sp . . . bank
bandar, Mal., Per. . . . town, port, harbor
bang, Siam . . . village
bassin, Fr . . . basin
batang, Indon., Mal . . . river
ben, Celt . . . mountain, summit
bender, Arab . . . harbor, port
bereg, Rus . . . coast, shore
berg, Du., Ger., Nor., Swe. . . . mountain, hill
bir, Arab . . . well
birkat, Arab . . . lake, pond, pool
bit, Arab . . . house
bjaerg, Dan., Nor . . . mountain
bocche, It . . . mouth
boğazı, Tur . . . strait
bois, Fr . . . forest, wood
boloto, Rus . . . marsh
bolsón, Sp. . . . flat-floored desert valley
boreal, Sp . . . northern
borg, Dan., Nor., Swe . . . castle, town
borgo, It . . . town, suburb
bosch, Du . . . forest, wood
bouche, Fr . . . river mouth
bourg, Fr . . . town, borough
bro, Dan., Nor., Swe . . . bridge
brücke, Ger . . . bridge
bucht, Ger . . . bay, bight
bugt, Dan., Nor., Swe . . . bay, gulf
bulu, Indon . . . mountain
burg, Du., Ger . . . castle, town
buri, Siam . . . town
burun, burnu, Tur . . . cape
by, Dan., Nor., Swe . . . village
caatinga, Port. (Brazil) . . . open brushland
cabezo, Sp . . . summit
cabo, Port., Sp . . . cape
campo, It., Port., Sp . . . plain, field
campos, Port. (Brazil) . . . plains
cañón, Sp . . . canyon
cap, Fr . . . cape
capo, It . . . cape
casa, It., Port., Sp . . . house
castello, It., Port . . . castle, fort
castillo, Sp . . . castle
càte, Fr . . . hill
çay, Tur . . . stream, river
cayo, Sp . . . rock, shoal, islet
cerro, Sp . . . mountain, hill
champ, Fr . . . field
chang, Chn . . . village, middle
château, Fr . . . castle
chen, Chn . . . market town
chiang, Chn . . . river
chott, Arab . . . salt lake
chou, Chn. . . . capital of district; island
chu, Tib . . . water, stream
cidade, Port . . . town, city
cima, Sp . . . summit, peak
città, It . . . town, city
ciudad, Sp . . . town, city
cochilha, Port . . . ridge
col, Fr . . . pass
colina, Sp . . . hill
cordillera, Sp . . . mountain chain
costa, It., Port., Sp . . . coast
côte, Fr . . . coast
cuchilla, Sp . . . mountain ridge
dağ, Tur . . . mountain(s)
dake, Jap . . . peak, summit
dal, Dan., Du., Nor., Swe . . . valley
dan, Kor . . . point, cape
danau, Indon . . . lake
dar, Arab . . . house, abode, country
darya, Per . . . river, sea
dasht, Per . . . plain, desert
deniz, Tur . . . sea
désert, Fr . . . desert
deserto, It . . . desert
desierto, Sp . . . desert
détroit, Fr . . . strait
dijk, Du . . . dam, dike
djebel, Arab . . . mountain
do, Kor . . . island
dorf, Ger . . . village
dorp, Du . . . village
duin, Du . . . dune
dzong, Tib. . . . fort, administrative capital
eau, Fr . . . water
ecuador, Sp . . . equator
eiland, Du . . . island
elv, Dan., Nor . . . river, stream
embalse, Sp . . . reservoir
erg, Arab . . . dune, sandy desert
est, Fr., It . . . east
estado, Sp . . . state
este, Port., Sp . . . east
estrecho, Sp . . . strait
étang, Fr . . . pond, lake
état, Fr . . . state
eyjar, Ice . . . islands
feld, Ger . . . field, plain
festung, Ger . . . fortress
fiume, It . . . river
fjäll, Swe . . . mountain
fjärd, Swe . . . bay, inlet
fjeld, Nor . . . mountain, hill
fjord, Dan., Nor . . . fiord, inlet
fjördur, Ice . . . fiord, inlet
fleuve, Fr . . . river
flod, Dan., Swe . . . river
flói, Ice . . . bay, marshland
fluss, Ger . . . river
foce, It . . . river mouth
fontein, Du . . . a spring
forêt, Fr . . . forest
fors, Swe . . . waterfall
forst, Ger . . . forest
fos, Dan., Nor . . . waterfall
fu, Chn . . . town, residence
fuente, Sp . . . spring, fountain
fuerte, Sp . . . fort
furt, Ger . . . ford
gang, Kor . . . stream, river
gangri, Tib . . . mountain
gat, Dan., Nor . . . channel
gàve, Fr . . . stream
gawa, Jap . . . river
gebergte, Du . . . mountain range
gebiet, Ger . . . district, territory
gebirge, Ger . . . mountains
ghat, India . . . pass, mountain range
gobi, Mong . . . desert
gol, Mong . . . river
göl, gölü, Tur . . . lake
golf, Du., Ger . . . gulf, bay
golfe, Fr . . . gulf, bay
golfo, It., Port., Sp . . . gulf, bay
gomba, gompa, Tib . . . monastery
gora, Rus., So. Slav . . . mountain
góra, Pol . . . mountain
gorod, Rus . . . town
grad, Rus., So. Slav . . . town
guba, Rus . . . bay, gulf
gundung, Indon . . . mountain
guntō, Jap . . . archipelago
gunung, Mal . . . mountain
haf, Swe . . . sea, ocean
hafen, Ger . . . port, harbor
haff, Ger . . . gulf, inland sea
hai, Chn . . . sea, lake
hama, Jap . . . beach, shore
hamada, Arab . . . rocky plateau
hamn, Swe . . . harbor
hāmūn, Per . . . swampy lake, plain
hantō, Jap . . . peninsula
hassi, Arab . . . well, spring
haus, Ger . . . house
haut, Fr . . . summit, top
hav, Dan., Nor . . . sea, ocean
havn, Dan., Nor . . . harbor, port
havre, Fr . . . harbor, port
háza, Hung . . . house, dwelling of
heim, Ger . . . hamlet, home
hem, Swe . . . hamlet, home
higashi, Jap . . . east
hisar, Tur . . . fortress
hissar, Arab . . . fort
ho, Chn . . . river
hoek, Du . . . cape
hof, Ger . . . court, farmhouse
höfn, Ice . . . harbor
hoku, Jap . . . north
holm, Dan., Nor., Swe . . . island
hora, Czech . . . mountain
horn, Ger . . . peak
hoved, Dan., Nor . . . cape
hsien, Chn . . . district, district capital
hu, Chn . . . lake
hügel, Ger . . . hill
huk, Dan., Swe . . . point
hus, Dan., Nor., Swe . . . house
île, Fr . . . island
ilha, Port . . . island
indsö, Dan., Nor . . . lake
insel, Ger . . . island
insjö, Swe . . . lake
irmak, irmagi, Tur . . . river
isla, Sp . . . island
isola, It . . . island
istmo, It., Sp . . . isthmus
järvi, jaur, Fin . . . lake
jebel, Arab . . . mountain
jima, Jap . . . island
jökel, Nor . . . glacier
joki, Fin . . . river
jökull, Ice . . . glacier
kaap, Du . . . cape
kai, Jap . . . bay, gulf, sea
kaikyō, Jap . . . channel, strait
kalat, Per . . . castle, fortress
kale, Tur . . . fort
kali, Mal . . . creek, river
kand, Per . . . village
kang, Chn . . . mountain ridge; village
kap, Dan., Ger . . . cape
kapp, Nor., Swe . . . cape
kasr, Arab . . . fort, castle
kawa, Jap . . . river
kefr, Arab . . . village
kei, Jap . . . creek, river
ken, Jap . . . prefecture
khor, Arab . . . bay, inlet
khrebet, Rus . . . mountain range
kiang, Chn . . . large river
king, Chn . . . capital city, town
kita, Jap . . . north
ko, Jap . . . lake
köbstad, Dan . . . market-town
kol, Mong . . . lake
kólpos, Gr . . . gulf
kong, Chn . . . river
kopf, Ger . . . head, summit, peak
köpstad, Swe . . . market-town
körfezi, Tur . . . gulf
kosa, Rus . . . spit
kou, Chn . . . river mouth
köy, Tur . . . village
kraal, Du. (Africa) . . . native village
ksar, Arab . . . fortified village
kuala, Mal . . . bay, river mouth
kuh, Per . . . mountain
kum, Tur . . . sand
kuppe, Ger . . . summit
küste, Ger . . . coast
kyo, Jap . . . town, capital
la, Tib . . . mountain pass
labuan, Mal . . . anchorage, port
lac, Fr . . . lake
lago, It., Port., Sp . . . lake
lagoa, Port . . . lake, marsh
laguna, It., Port., Sp . . . lagoon, lake
lahti, Fin . . . bay, gulf
län, Swe . . . county
landsby, Dan., Nor . . . village
liehtao, Chn . . . archipelago
liman, Tur . . . bay, port
ling, Chn . . . pass, ridge, mountain
llanos, Sp . . . plains
loch, Celt. (Scotland) . . . lake, bay
loma, Sp . . . long, low hill
lough, Celt. (Ireland) . . . lake, bay
machi, Jap . . . town
man, Kor . . . bay
mar, Port., Sp . . . sea
mare, It., Rom . . . sea
marisma, Sp . . . marsh, swamp
mark, Ger . . . boundary, limit
massif, Fr . . . block of mountains
mato, Port . . . forest, thicket
me, Siam . . . river
meer, Du., Ger . . . lake, sea
mer, Fr . . . sea
mesa, Sp . . . flat-topped mountain
meseta, Sp . . . plateau
mina, Port., Sp . . . mine
minami, Jap . . . south
minato, Jap . . . harbor, haven
misaki, Jap . . . cape, headland
mont, Fr . . . mount, mountain
montagna, It . . . mountain
montagne, Fr . . . mountain
montaña, Sp . . . mountain
monte, It., Port., Sp. . . . mount, mountain
more, Rus., So. Slav . . . sea
morro, Port., Sp . . . hill, bluff
mühle, Ger . . . mill
mund, Ger . . . mouth, opening
mündung, Ger . . . river mouth
mura, Jap . . . township
myit, Bur . . . river
mys, Rus . . . cape
nada, Jap . . . sea
nadi, India . . . river, creek
naes, Dan., Nor . . . cape
nafud, Arab . . . desert of sand dunes
nagar, India . . . town, city
nahr, Arab . . . river
nam, Siam . . . river, water
nan, Chn., Jap . . . south
näs, Nor., Swe . . . cape
nez, Fr . . . point, cape
nishi, nisi, Jap . . . west
njarga, Fin . . . peninsula
nong, Siam . . . marsh
noord, Du . . . north
nor, Mong . . . lake
nord, Dan., Fr., Ger., It., Nor., Swe . . . north
norte, Port., Sp . . . north
nos, Rus . . . cape
nyasa, Bantu . . . lake
ö, Dan., Nor., Swe . . . island
occidental, Sp . . . western
ocna, Rom . . . salt mine
odde, Dan., Nor . . . point, cape
oeste, Port., Sp . . . west
oka, Jap . . . hill
oost, Du . . . east
oriental, Sp . . . eastern
óros, Gr . . . mountain
ost, Ger., Swe . . . east
öster, Dan., Nor., Swe . . . eastern
ostrov, Rus . . . island
oued, Arab . . . river, stream
ouest, Fr . . . west
ozero, Rus . . . lake
pää, Fin . . . mountain
padang, Mal . . . plain, field
pampas, Sp. (Argentina) . . . grassy plains
pará, Indian (Brazil) . . . river
pas, Fr . . . channel, passage
paso, Sp . . . mountain pass, passage
passo, It., Port. mountain pass, passage, strait
patam, India . . . city, town
pei, Chn . . . north
pélagos, Gr . . . open sea
pegunungan, Indon . . . mountains
peña, Sp . . . rock
peresheyek, Rus . . . isthmus
pertuis, Fr . . . strait
peski, Rus . . . desert
pic, Fr . . . mountain peak
pico, Port., Sp . . . mountain peak
piedra, Sp . . . stone, rock
ping, Chn . . . plain, flat
planalto, Port . . . plateau
planina, Yugo . . . mountains
playa, Sp . . . shore, beach
pnom, Camb . . . mountain
pointe, Fr . . . point
polder, Du., Ger . . . reclaimed marsh
polje, So. Slav . . . plain, field
poluostrov, Rus . . . peninsula
pont, Fr . . . bridge
ponta, Port . . . point, headland
ponte, It., Port . . . bridge
pore, India . . . city, town
porthmós, Gr . . . strait
porto, It., Port . . . port, harbor
potamós, Gr . . . river
p'ov, Rus . . . peninsula
prado, Sp . . . field, meadow
presqu'île, Fr . . . peninsula
proliv, Rus . . . strait
pu, Chn . . . commercial village
pueblo, Sp . . . town, village
puerto, Sp . . . port, harbor
pulau, Indon . . . island
punkt, Ger . . . point
punt, Du . . . point
punta, It., Sp . . . point
pur, India . . . city, town
puy, Fr . . . peak
qal'a, qal'at, Arab . . . fort, village
qasr, Arab . . . fort, castle
rann, India . . . wasteland
ra's, Arab . . . cape, head
reka, Rus., So. Slav . . . river
represa, Port . . . reservoir
rettō, Jap . . . island chain
ría, Sp . . . estuary
ribeira, Port . . . stream
riberão, Port . . . river
rio, It., Port . . . stream, river
río, Sp . . . river
rivière, Fr . . . river
roca, Sp . . . rock
rt, Yugo . . . cape
rūd, Per . . . river
saari, Fin . . . island
sable, Fr . . . sand
sahara, Arab . . . desert, plain
saki, Jap . . . cape
sal, Sp . . . salt
salar, Sp . . . salt flat, salt la
salto, Sp . . . waterf
san, Jap., Kor . . . mountain,
sat, satul, Rom . . . villa
schloss, Ger . . . cas
sebkha, Arab . . . salt mar
see, Ger . . . lake, s
şehir, Tur . . . town, c
selat, Indon . . . strea
selvas, Port. (Brazil) . . . tropical rain fore
seno, Sp . . . b
serra, Port . . . mountain cha
serranía, Sp . . . mountain rid
seto, Jap . . . str
severnaya, Rus . . . northe
shahr, Per . . . town, c
shan, Chn . . . mountain, hill, isla
shatt, Arab . . . riv
shi, Jap . . . c
shima, Jap . . . isla
shōtō, Jap . . . archipela
si, Chn . . . west, weste
sierra, Sp . . . mountain ran
sjö, Nor., Swe . . . lake, s
sö, Dan., Nor . . . lake, s
söder, södra, Swe . . . sou
song, Annam . . . riv
sopka, Rus . . . peak, volca
source, Fr . . . a spri
spitze, Ger . . . summit, po
staat, Ger . . . sta
stad, Dan., Du., Nor., Swe. . . . city, tov
stadt, Ger . . . city, tov
stato, It . . . sta
step', Rus . . . treeless plain, step
straat, Du . . . str
strand, Dan., Du., Ger., Nor., Swe . . . shore, bea
stretto, It . . . str
strom, Ger . . . river, strea
ström, Dan., Nor., Swe. . . . stream, riv
stroom, Du . . . stream, riv
su, suyu, Tur . . . water, riv
sud, Fr., Sp . . . sou
süd, Ger . . . sou
suidō, Jap . . . chann
sul, Port . . . sou
sund, Dan., Nor., Swe . . . sour
sungai, sungei, Indon., Mal . . . riv
sur, Sp . . . sou
syd, Dan., Nor., Swe . . . sou
tafelland, Ger . . . platea
take, Jap . . . peak, summ
tal, Ger . . . valle
tanjung, tanjong, Mal . . . ca
tao, Chn . . . islar
târg, târgul, Rom . . . market, tow
tell, Arab . . . h
teluk, Indon . . . bay, gu
terra, It . . . lar
terre, Fr . . . earth, lar
thal, Ger . . . valle
tierra, Sp . . . earth, lar
tō, Jap . . . east; islar
tonle, Camb . . . river, la
top, Du . . . pea
torp, Swe . . . hamlet, cottag
tsangpo, Tib . . . riv
tsi, Chn . . . village, boroug
tso, Tib . . . la
tsu, Jap . . . harbor, po
tundra, Rus . . . treeless arctic plair
tung, Chn . . . ea
tuz, Tur . . . sa
udde, Swe . . . cap
ufer, Ger . . . shore, riverbar
ujung, Indon . . . point, cap
umi, Jap . . . sea, gu
ura, Jap . . . bay, coast, cree
ust'ye, Rus . . . river mou
valle, It., Port., Sp . . . valle
vallée, Fr . . . valle
valli, It . . . la
vár, Hung . . . fortres
város, Hung . . . tow
varoš, So. Slav . . . tow
veld, Du . . . open plain, fiel
verkh, Rus . . . top, summ
ves, Czech . . . villag
vest, Dan., Nor., Swe . . . wes
vik, Swe . . . cove, ba
vila, Port . . . tow
villa, Sp . . . tow
villar, Sp . . . village, hamle
ville, Fr . . . town, cit
vostok, Rus . . . eas
wad, wādī, Arab. . . . intermittent strear
wald, Ger . . . forest, woodlan
wan, Chn., Jap . . . bay, gu
weiler, Ger . . . hamlet, villag
westersch, Du . . . wester
wüste, Ger . . . deser
yama, Jap . . . mountai
yarimada, Tur . . . peninsul
yug, Rus . . . sout
zaki, Jap . . . cap
zaliv, Rus . . . bay, gu
zapad, Rus . . . wes
zee, Du . . . se
zemlya, Rus . . . lan
zuid, Du . . . sout

BBREVIATIONS OF GEOGRAPHICAL
AMES AND TERMS

PRONUNCIATION OF GEOGRAPHICAL NAMES

....	Afghanistan
....	Africa
S.	Alaska, U.S.
S.	Alabama, U.S.
....	Albania
....	Algeria
am.	American Samoa
....	Andorra
....	Angola
....	Antarctica
....	Antigua and Barbuda
....	Aqueduct
S.	Arkansas, U.S.
....	Argentina
....	Armenia
....	Airport
....	Austria
....	Australia
S.	Arizona, U.S.
....	Azerbaijan
....	Bay, Gulf, Inlet, Lagoon
....	Bahamas
....	Bahrain
....	Barbados
....	Burundi
....	Belgium
....	Belarus
....	Bermuda
....	Bhutan
....	Undersea Bank
....	Building
....	Bulgaria
....	Bangladesh
....	Bolivia
....	Bosnia and Herzegovina
....	Botswana
....	Brazil
....	Brunei
r. Is.	British Virgin Islands
....	Bight
na	Burkina Faso
....	Cape, Point
.S.	California, U.S.
....	Cameroon
.	Cambodia
....	Canal
....	Canada
....	Central African Republic
Is.	Cayman Islands
....	Cote d'Ivoire
....	Cliff, Escarpment
....	County, Parish
J.S.	Colorado, U.S.
....	Colombia
....	Comoros
....	Continent
Is.	Cook Islands
....	Costa Rica
....	Croatia
....	Coast, Beach
.S.	Connecticut, U.S.
....	Cape Verde
....	Cyprus
h Rep.	Czech Republic
....	Delta
U.S.	District of Columbia, U.S.
.U.S.	Delaware, U.S.
....	Denmark
....	Dependency, Colony
....	Depression
....	Department, District
....	Desert
....	Djibouti
....	Dominica
. Rep.	Dominican Republic
C.	Democratic Republic of the Congo
....	Ecuador
.	Educational Facility
al.	El Salvador
.U.K.	England, U.K.
Gui.	Equatorial Guinea
....	Eritrea
....	Estonia
....	Estuary
....	Ethiopia
mor	East Timor
....	Europe
. Is.	Falkland Islands
ds.	Faroe Islands
....	Finland
....	Fjord
.U.S.	Florida, U.S.
....	Forest, Moor
....	France
u.	French Guiana
oly.	French Polynesia
.U.S.	Georgia, U.S.
.	The Gambia
....	Gaza Strip
....	Georgia
....	Germany

Grc.	Greece
Gren.	Grenada
Grnld.	Greenland
Guad.	Guadeloupe
Guat.	Guatemala
Guern.	Guernsey
Gui.	Guinea
Gui.-B.	Guinea-Bissau
Guy.	Guyana
Hi., U.S.	Hawaii, U.S.
hist.	Historic Site, Ruins
hist. reg.	Historic Region
Hond.	Honduras
Hung.	Hungary
i.	Island
Ia., U.S.	Iowa, U.S.
ice	Ice Feature, Glacier
Ice.	Iceland
Id., U.S.	Idaho, U.S.
Il., U.S.	Illinois, U.S.
In., U.S.	Indiana, U.S.
Indon.	Indonesia
I. of Man	Isle of Man
I.R.	Indian Reservation
Ire.	Ireland
is.	Islands
Isr.	Israel
isth.	Isthmus
Jam.	Jamaica
Jord.	Jordan
Kaz.	Kazakhstan
Kir.	Kiribati
Kor., N.	Korea, North
Kor., S.	Korea, South
Ks., U.S.	Kansas, U.S.
Kuw.	Kuwait
Ky., U.S.	Kentucky, U.S.
Kyrg.	Kyrgyzstan
l.	Lake, Pond
La., U.S.	Louisiana, U.S.
Lat.	Latvia
Leb.	Lebanon
Leso.	Lesotho
Lib.	Liberia
Liech.	Liechtenstein
Lith.	Lithuania
Lux.	Luxembourg
Ma., U.S.	Massachusetts, U.S.
Mac.	Macedonia
Madag.	Madagascar
Malay.	Malaysia
Mald.	Maldives
Marsh. Is.	Marshall Islands
Mart.	Martinique
Maur.	Mauritania
May.	Mayotte
Md., U.S.	Maryland, U.S.
Me., U.S.	Maine, U.S.
Mex.	Mexico
Mi., U.S.	Michigan, U.S.
Micron.	Micronesia, Federated States of
Mn., U.S.	Minnesota, U.S.
Mo., U.S.	Missouri, U.S.
Mol.	Moldova
Mong.	Mongolia
Monts.	Montserrat
Mor.	Morocco
Moz.	Mozambique
Ms., U.S.	Mississippi, U.S.
Mt., U.S.	Montana, U.S.
mth.	River Mouth or Channel
mtn.	Mountain
mts.	Mountains
Mwi.	Malawi
Mya.	Myanmar
N.A.	North America
N.C., U.S.	North Carolina, U.S.
N. Cal.	New Caledonia
N.D., U.S.	North Dakota, U.S.
Ne., U.S.	Nebraska, U.S.
neigh.	Neighborhood
Neth.	Netherlands
Neth. Ant.	Netherlands Antilles
N.H., U.S.	New Hampshire, U.S.
Nic.	Nicaragua
Nig.	Nigeria
N. Ire., U.K.	Northern Ireland, U.K.
N.J., U.S.	New Jersey, U.S.
N.M., U.S.	New Mexico, U.S.
N. Mar. Is.	Northern Mariana Islands
Nmb.	Namibia
Nor.	Norway
Nv., U.S.	Nevada, U.S.
N.Y., U.S.	New York, U.S.
N.Z.	New Zealand
o.	Ocean
Oc.	Oceania
Oh., U.S.	Ohio, U.S.

Ok., U.S.	Oklahoma, U.S.
Or., U.S.	Oregon, U.S.
p.	Pass
Pa., U.S.	Pennsylvania, U.S.
Pak.	Pakistan
Pan.	Panama
Pap. N. Gui.	Papua New Guinea
Para.	Paraguay
pen.	Peninsula
Phil.	Philippines
Pit.	Pitcairn
pl.	Plain, Flat
plat.	Plateau, Highland
Pol.	Poland
Port.	Portugal
P.R.	Puerto Rico
prov.	Province, Region
pt. of i.	Point of Interest
r.	River, Creek
Reu.	Reunion
rec.	Recreational Site, Park
reg.	Physical Region
rel.	Religious Institution
res.	Reservoir
rf.	Reef, Shoal
R.I., U.S.	Rhode Island, U.S.
Rom.	Romania
Rw.	Rwanda
S.A.	South America
S. Afr.	South Africa
Sau. Ar.	Saudi Arabia
S.C., U.S.	South Carolina, U.S.
sci.	Scientific Station
Scot., U.K.	Scotland, U.K.
S.D., U.S.	South Dakota, U.S.
sea feat.	Undersea Feature
Sen.	Senegal
Serb.	Serbia and Montenegro
Sey.	Seychelles
S. Geor.	South Georgia
Sing.	Singapore
S.L.	Sierra Leone
Slvk.	Slovakia
Slvn.	Slovenia
S. Mar.	San Marino
Sol. Is.	Solomon Islands
Som.	Somalia
Sp. N. Afr.	Spanish North Africa
Sri L.	Sri Lanka
St. Hel.	St. Helena
St. K./N.	St. Kitts and Nevis
St. Luc.	St. Lucia
St. P./M.	St. Pierre and Miquelon
strt.	Strait, Channel, Sound
S. Tom./P.	Sao Tome and Principe
St. Vin.	St. Vincent and the Grenadines
Sur.	Suriname
Sval.	Svalbard
sw.	Swamp, Marsh
Swaz.	Swaziland
Swe.	Sweden
Switz.	Switzerland
Tai.	Taiwan
Taj.	Tajikistan
Tan.	Tanzania
T./C. Is.	Turks and Caicos Islands
ter.	Territory
Thai.	Thailand
Tn., U.S.	Tennessee, U.S.
trans.	Transportation Facility
Trin.	Trinidad and Tobago
Tun.	Tunisia
Tur.	Turkey
Turkmen.	Turkmenistan
Tx., U.S.	Texas, U.S.
U.A.E.	United Arab Emirates
Ug.	Uganda
U.K.	United Kingdom
Ukr.	Ukraine
Ur.	Uruguay
U.S.	United States
Ut., U.S.	Utah, U.S.
Uzb.	Uzbekistan
Va., U.S.	Virginia, U.S.
val.	Valley, Watercourse
Ven.	Venezuela
Viet.	Vietnam
V.I.U.S.	Virgin Islands (U.S.)
vol.	Volcano
Vt., U.S.	Vermont, U.S.
Wa., U.S.	Washington, U.S.
W.B.	West Bank
Wi., U.S.	Wisconsin, U.S.
W. Sah.	Western Sahara
wtfl.	Waterfall
W.V., U.S.	West Virginia, U.S.
Wy., U.S.	Wyoming, U.S.
Zam.	Zambia
Zimb.	Zimbabwe

Key to the Sound Values of Letters and Symbols Used in the Index to Indicate Pronunciation

ă-ăt; băttle
ă-fin*ă*l; appe*ă*l
ā-rāte; elāte
å-senåte; inanimåte
ä-ärm; cälm
à-àsk; bàth
à-sof*à*; m*à*rine (short neutral or indeterminate sound)
â-fâre; prepâre
ch-choose; church
dh-as th in other; either
ē-bē; ēve
ĕ-ĕvent; crĕate
ĕ-bĕt; ĕnd
ĕ-recĕnt (short neutral or indeterminate sound)
ẽ-cratẽr; cindẽr
g-gō; gāme
gh-guttural g
ĭ-bĭt; wĭll
ĭ-(short neutral or indeterminate sound)
ī-rīde; bīte
к-gutteral k as *ch* in German *ich*
ng-sing
ŋ-baŋk; liŋger
и-indicates nasalized
ŏ-nŏd; ŏdd
ŏ-cŏmmit; cŏnnect
ō-ōld; bōld
ô-ôbey; hôtel
ô-ôrder; nôrth
oi-boil
o͞o-fo͞od; ro͞ot
ò-as oo in foot; wood
ou-out; thou
s-soft; so; sane
sh-dish; finish
th-thin; thick
ū-pūre; cūre
ů-ůnite; ůsûrp
û-ûrn; fûr
ŭ-stŭd; ŭp
ŭ-circ*ŭ*s; s*ŭ*bmit
ü-as in French *tu*
zh-as *z* in azure
'-indeterminate vowel sound

In many cases the spelling of foreign geographical names does not even remotely indicate the pronunciation to an American, i.e., Słupsk in Poland is pronounced swôpsk; Jujuy in Argentina is pronounced ho͞ohwē'; La Spezia in Italy is lä-spĕ'zyä.

This condition is hardly surprising, however, when we consider that in our own language Worcester, Massachusetts, is pronounced wôs'tẽr; Sioux City, Iowa, so͞o sĭ'tĭ; Schuylkill Haven, Pennsylvania, sko͞ol'kĭl hā-vĕn; Poughkeepsie, New York, pŏ-kĭp'sĕ.

The indication of pronunciation of geographic names presents several peculiar problems:

1. Many foreign tongues use sounds that are not present in the English language and which an American cannot normally articulate. Thus, though the nearest English equivalent sound has been indicated, only approximate results are possible.

2. There are several dialects in each foreign tongue which cause variation in the local pronunciation of names. This also occurs in identical names in the various divisions of a great language group, as the Slavic or the Latin.

3. Within the United States there are marked differences in pronunciation, not only of local geographic names, but also of common words, indicating that the sound and tone values for letters as well as the placing of the emphasis vary considerably from one part of the country to another.

4. A number of different letters and diacritical combinations could be used to indicate essentially the same or approximate pronunciations.

Some variation in pronunciation other than that indicated in this index may be encountered, but such a difference does not necessarily indicate that either is in error, and in many cases it is a matter of individual choice as to which is preferred. In fact, an exact indication of pronunciation of many foreign names using English letters and diacritical marks is extremely difficult and sometimes impossible.

PRONOUNCING INDEX

his universal index includes in a single alphabetical list approximately 1,000 ames of features that appear on the reference maps. Each name is followed y a page reference and geographical coordinates.

bbreviation and Capitalization Abbreviations of names on the maps have been tandardized as much as possible. Names that are abbreviated on the maps are enerally spelled out in full in the index. Periods are used after all abbreviations egardless of local practice. The abbreviation "St." is used only for "Saint". Sankt" and other forms of this term are spelled out.

lost initial letters of names are capitalized, except for a few Dutch names, uch as "s-Gravenhage". Capitalization of noninitial words in a name generally ollows local practice.

lphabetization Names are alphabetized in the order of the letters of the English lphabet. Spanish ll and ch, for example, are not treated as direct letters. urthermore, diacritical marks are disregarded in alphabetization — German r Scandinavian ä or ö are treated as a or o.

he names of physical features may appear inverted, since they are always lphabetized under the proper, not the generic, part of the name, thus: "Gibraltar, trait of". Otherwise every entry, whether consisting of one word or more, is lphabetized as a single continuous entity. "Lakeland", for example, appears fter "La Crosse" and before "La Salle". Names beginning with articles (Le arve, Den Helder, Al Manāmah, Ad Dawhah) are not inverted.

In the case of identical names, towns are listed first, then political divisions, then physical features.

Generic Terms Except for cities, the names of all features are followed by terms that represent broad classes of features, for example, Mississippi, r. or Alabama, state. A list of all abbreviations used in the index is on page 85.

Country names and the names of features that extend beyond the boundaries of one county are followed by the name of the continent in which each is located. Country designations follow the names of all other places in the index. The locations of places in the United States and the United Kingdom are further defined by abbreviations that include the state or political division in which each is located.

Pronunciations Pronunciations are included for most names listed. An explanation of the pronunciation system used appears on page 85.

Page References and Geographical Coordinates The geographical coordinates and page references are found in the last columns of each entry.

If a page contains several maps or insets, a lowercase letter identifies the specific map or inset.

Latitude and longitude coordinates for point features, such as cities and mountain peaks, indicate the location of the symbols. For extensive areal features, such as countries or mountain ranges, or linear features, such as canals and rivers, locations are given for the position of the type as it appears on the map.

LACE (Pronunciation)	PAGE	LAT.	LONG.
A			
bidjan, C. Iv. (ä-bēd-zhän´)	6	5°19´N	4°02´W
cadia National Park, rec., Me., U.S. (ā-kā´dǐ-á)	59	44°19´N	68°01´W
concagua, Cerro, mtn., Arg. (ä-kôn-kä´gwä)	4	32°38´S	70°00´W
çores (Azores), is., Port.	2	37°44´N	29°25´W
ddis Ababa, Eth. (ä-kôn-kä´gwä)	3	9°00´N	38°44´E
delaide, Austl. (ăd´ĕ-lād)	3	34°46´S	139°08´E
den (´Adan), Yemen (ä´dĕn)	3	12°48´N	45°00´E
den, Gulf of, b.	5	11°45´N	45°45´E
driatic Sea, sea, Eur.	5	43°30´N	14°27´E
egean Sea, sea, Eur. (ē-jē´ăn)	5	39°04´N	24°56´E
fghanistan, nation, Asia (ăf-găn-ĭ-stăn´)	3	33°00´N	63°00´E
frica, cont.	68	10°00´N	22°00´E
gulhas, Cape, c., S. Afr. (ä-gōōl´yäs)	5	34°47´S	20°00´E
hmadnagar, India (ä´mŭd-nŭ-gŭr)	7	19°09´N	74°45´E
kron, Oh., U.S. (ăk´rŭn)	53	41°05´N	81°30´W
laska, state, U.S. (á-lăs´ká)	60	64°00´N	150°00´W
laska, Gulf of, b., Ak., U.S. (á-lăs´ká)	61	57°42´N	147°40´W
laska Peninsula, pen., Ak., U.S. (á-lăs´ká)	60	55°50´N	162°10´W
laska Range, mts., Ak., U.S. (á-lăs´ká)	61	62°00´N	152°18´W
lbany, N.Y., U.S. (ôl´bá-nǐ)	43	42°40´N	73°50´W
lbuquerque, N.M., U.S. (ăl-bû-kûr´kê)	42	35°05´N	106°40´W
ldabra Islands, is., Sey. (ăl-dä´brä)	5	9°16´S	46°17´E
leutian Islands, is., Ak., U.S. (á-lu´shăn)	4	52°40´N	177°30´W
leutian Trench, deep	4	50°40´N	177°10´E
lexander Island, i., Ant. (ăl-ĕg-zăn´dĕr)	4	71°00´S	71°00´W
lexandria, Egypt (ăl-ĕg-zăn´drǐ-á)	6	31°12´N	29°58´E
lgeria, nation, Afr. (ăl-gē´rǐ-á)	68	28°45´N	1°00´E
lgiers, Alg. (ăl-jērs)	3	36°51´N	2°56´E
llegany Indian Reservation, I.R., N.Y., U.S. (ăl-ê-gā´nǐ)	59	42°05´N	78°55´W
llentown, Pa., U.S. (ălĕn-toun)	53	40°35´N	75°30´W
lps, mts., Eur. (ălps)	5	46°18´N	8°42´E
ltai Mountains, mts., Asia (ăl´tī´)	5	49°11´N	87°15´E
mazon (Amazonas) (Solimões), r., S.A.	4	2°03´S	53°18´W
mbre, Cap d´, c., Madag.	5	12°06´S	49°15´E
merican Samoa, dep., Oc.	2	14°20´S	170°00´W
mirante Islands, is., Sey.	5	6°02´S	52°30´E
msterdam, Île, i., Afr.	5	37°52´S	77°32´E
mu Darya, r., Asia (ä-mô-dä´rēä)	5	38°30´N	64°00´E
mur, r., Asia	5	49°00´N	136°00´E
nchorage, Ak., U.S. (ăŋ´kĕr-åj)	60	61°12´N	149°48´W

PLACE (Pronunciation)	PAGE	LAT.	LONG.
Andaman Islands, is., India (ăn-dá-măn´)	5	11°38´N	92°17´E
Andes Mountains, mts., S.A. (ăn´dēz) (ăn´dās)	4	13°00´S	75°00´W
Angmagssalik, Grnld. (äŋ-má´sá-lĭk)	60	65°40´N	37°40´W
Angola, nation, Afr. (ăŋ-gō´lá)	68	14°15´S	16°00´E
Anguilla, dep., N.A.	66	18°15´N	62°54´W
Aniakchak National Monument, rec., Ak., U.S.	58a	56°50´N	157°50´W
Ankara, Tur. (äŋ´ká-rá)	3	39°55´N	32°50´E
An Nafūd, des., Sau. Ar.	5	28°30´N	40°30´E
Ann Arbor, Mi., U.S. (ăn är´bĕr)	53	42°15´N	83°45´W
Antananarivo, Madag.	3	18°51´S	47°40´E
Antarctica, cont.	3	80°15´S	127°00´E
Antarctic Peninsula, pen., Ant.	4	70°00´S	65°00´W
Antigua and Barbuda, nation, N.A.	66	17°15´N	61°15´W
Antofagasta, Chile (än-tô-fä-gäs´tä)	2	23°32´S	70°21´W
Aoraki (Cook, Mount), mtn., N.Z.	5	43°27´S	170°13´E
Appalachian Mountains, mts., N.A. (ăp-á-lăch´ǐ-án)	61	37°20´N	82°00´W
Appennino, mts., Italy (äp-pĕn-nē´nô)	5	43°48´N	11°06´E
Arabian Sea, sea (á-rā´bǐ-án)	5	16°00´N	65°15´E
Arafura Sea, sea (ä-rä-fōō´rä)	5	8°40´S	130°00´E
Aral Sea, sea, Asia	5	45°17´N	60°02´E
Ararat, Mount, mtn., Tur. (är´árát)	5	39°50´N	44°20´E
Arctic Ocean, o.	61	85°00´N	170°00´E
Argentina, nation, S.A. (är-jĕn-tē´ná)	2	35°30´S	67°00´W
Arkansas, r., U.S. (är´kăn-sô) (är-kăn´sás)	61	37°30´N	97°00´W
Arkhangelsk (Archangel), Russia (är-kän´gĕlsk)	3	64°30´N	40°25´E
Armenia, nation, Asia	3	41°00´N	44°39´E
Aruba, dep., N.A. (ä-rōō´bä)	66	12°29´N	70°00´W
Ascension, i., St. Hel. (á-sĕn´shŭn)	4	8°00´S	13°00´W
Asia, cont.	5	50°00´N	100°00´E
Asia Minor, reg., Tur. (ā´zhá)	5	38°18´N	31°18´E
Asunción, Para. (ä-sōōn-syōn´)	2	25°25´S	57°30´W
Athabasca, l., Can. (ăth-á-băs´ká)	61	59°04´N	109°10´W
Athabasca, r., Can.	61	57°30´N	112°00´W
Athens (Athína), Grc.	3	38°00´N	23°38´E
Atlanta, Ga., U.S. (ăt-lăn´tá)	60	33°45´N	84°23´W
Atlantic Ocean, o.	4	5°00´S	25°00´W
Atlas Mountains, mts., Afr. (ăt´lăs)	4	31°22´N	4°57´W
Auckland, N.Z. (ôk´lănd)	3	36°53´S	174°45´E
Auckland Islands, is., N.Z.	5	50°30´S	166°30´E
Austin, Tx., U.S. (ôs´tǐn)	43	30°15´N	97°42´W
Australia, nation, Oc.	3	25°00´S	135°00´E
Austria, nation, Eur. (ôs´trǐ-á)	3	47°15´N	11°53´E
Azerbaijan, nation, Asia	3	40°30´N	47°30´E
Azores see Açores, is., Port.	2	37°44´N	29°25´W

PLACE (Pronunciation)	PAGE	LAT.	LONG.
B			
Badlands National Park, rec., S.D., U.S. (băd´ lănds)	58	43°56´N	102°37´W
Baffin Bay, b., N.A. (băf´ĭn)	61	72°00´N	65°00´W
Baffin Island, i., Can.	60	67°20´N	71°00´W
Baghdād, Iraq (băgh-däd´) (băg´dăd)	3	33°14´N	44°22´E
Bahamas, N.A. (bá-hä´más)	66	26°15´N	76°00´W
Baja California, state, Mex. (bä-hä)	60	30°15´N	117°25´W
Baja California, pen., Mex.	4	28°00´N	113°30´W
Baker, i., Oc. (bā´kēr)	2	1°00´N	176°00´W
Bakersfield, Ca., U.S. (bā´kērz-fēld)	53	35°23´N	119°00´W
Baku (Bakı), Azer. (bá-kōō´)	7	40°28´N	49°45´E
Balleny Islands, is., Ant. (băl´ĕ nê)	5	67°00´S	164°00´E
Balqash köli, l., Kaz.	5	45°58´N	72°15´E
Baltic Sea, sea, Eur. (bôl´tĭk)	5	55°20´N	16°50´E
Baltimore, Md., U.S. (bôl´tǐ-môr)	60	39°20´N	76°38´W
Banda, Laut (Banda Sea), sea, Indon.	5	6°05´S	127°28´E
Bandeira, Pico da, mtn., Braz. (pē´kò dä bän dä´rä)	4	20°27´S	41°47´W
Banff National Park, rec., Can. (bănf)	58	51°38´N	116°22´W
Bangalore, India (băŋ´gá´lôr)	7	13°03´N	77°39´E
Bangkok, Thai.	3	13°50´N	100°29´E
Bangladesh, nation, Asia	3	24°15´N	90°00´E
Banks Island, i., Can. (băŋks)	60	73°00´N	123°00´W
Barbados, nation, N.A. (bär-bā´dōz)	66	13°30´N	59°00´W
Barbuda, i., Antig. (bär-bōō´dá)	66	17°45´N	61°15´W
Barcelona, Spain (bär-thä-lō´nä)	6	41°25´N	2°08´E
Barents Sea, sea, Eur. (bä´rĕnts)	5	72°14´N	37°28´E
Barrow, Ak., U.S. (băr´ō)	60	71°20´N	156°00´W
Bass Strait, strt., Austl. (băs)	5	39°40´S	145°40´E
Baton Rouge, La., U.S. (băt´ŭn rōōzh´)	53	30°28´N	91°10´W
Baykal, Ozero (Lake Baikal), l., Russia	5	53°00´N	109°28´E
Beaufort Sea, sea, N.A.	61	70°30´N	138°40´W
Beijing, China	3	39°55´N	116°23´E
Belarus, nation, Eur.	3	53°30´N	25°33´E
Belém, Braz. (bá-lĕn´)	2	1°18´S	48°27´W
Belgium, nation, Eur. (bĕl´jǐ-ŭm)	3	51°00´N	2°52´E
Belize, nation, N.A.	66	17°00´N	88°40´W
Belle Isle, Strait of, strt., Can.	61	51°35´N	56°30´W
Belmopan, Belize	66	17°15´N	88°47´W
Belo Horizonte, Braz. (bĕ´lòre-sô´n-tĕ)	2	19°54´S	43°56´W
Bengal, Bay of, b., Asia (bĕn-gôl´)	5	17°30´N	87°00´E
Benin, nation, Afr. (bĕn-ēn´)	68	8°00´N	2°00´E
Bering Sea, sea (bē´rǐng)	61	58°00´N	175°00´W
Bering Strait, strt.	61	64°50´N	169°50´W
Berlin, Ger. (bĕr-lēn´)	6	52°31´N	13°28´E
Bermuda, dep., N.A.	60	32°20´N	65°45´W
Bhutan, nation, Asia (bōō-tän´)	3	27°15´N	90°30´E
Big Bend National Park, rec., Tx., U.S. (bǐg bĕnd)	58	29°15´N	103°15´W
Bighorn Lake, res., Mt., U.S. (bǐg hôrn)	58	45°00´N	108°10´W

t; finăl; rāte; senâte; ärm; ásk; sofá; fâre; ch-choose; dh-as th in other; bē; ĕvent; bĕt; recĕnt; cratẽr; g-gō; gh-guttural g; bĭt; ĭ-short neutral; rīde; κ-guttural k as ch in German ich;

PLACE (Pronunciation)	PAGE	LAT.	LONG.
Billings, Mt., U.S. (bĭl′ĭngz)	42	45°47′N	108°29′W
Bioko (Fernando Póo), i., Eq. Gui.	5	3°35′N	7°45′E
Birmingham, Eng., U.K.	6	52°29′N	1°53′W
Birmingham, Al., U.S. (bûr′mĭng-häm)	60	33°31′N	86°49′W
Biscay, Bay of, b., Eur. (bĭs′kā′)	5	45°19′N	3°51′W
Bismarck, N.D., U.S. (bĭz′märk)	42	46°48′N	100°46′W
Blackfeet Indian Reservation, I.R., Mt., U.S.	58	48°40′N	113°00′W
Blackfoot Indian Reserve, I.R., Can. (blăk′fo͝ot)	58	50°45′N	113°00′W
Black Hills, mts., U.S.	58	44°08′N	103°47′W
Black Sea, sea	5	43°01′N	32°16′E
Blanc, Mont, mtn., Eur. (môn blän)	5	45°50′N	6°53′E
Blood Indian Reserve, I.R., Can.	58	49°30′N	113°10′W
Bogotá, Col.	2	4°36′N	74°05′W
Boise, r., Id., U.S. (boi′zē)	43	43°43′N	116°30′W
Bolivia, nation, S.A. (bô-lĭv′ĭ-à)	2	17°00′S	64°00′W
Bonin Islands, is., Japan (bō′nĭn)	5	26°30′N	141°00′E
Booker T. Washington National Monument, rec., Va., U.S. (bo͝ok′ẽr tē wŏsh′ĭng-tŭn)	59	37°07′N	79°45′W
Boothia Peninsula, pen., Can. (bo͞o′thĭ-à)	60	73°30′N	95°00′W
Borneo, i., Asia	3	0°25′N	112°39′E
Bosnia and Herzegovina, nation, Eur.	3	44°15′N	17°30′E
Boston, Ma., U.S. (bôs′tŭn)	60	42°15′N	71°07′W
Bothnia, Gulf of, b., Eur. (bŏth′nĭ-à)	5	63°40′N	21°30′E
Botswana, nation, Afr. (bŏtswănă)	68	22°10′S	23°13′E
Bounty Islands, is., N.Z.	5	47°42′S	179°05′E
Bouvetøya, i., Ant.	5	55°00′S	3°00′E
Brasília, Braz. (brä-sē′lvä)	2	15°49′S	47°39′W
Brazil, nation, S.A. (brȧ-zĭl′)	2	9°00′S	53°00′W
Brazilian Highlands, mts., Braz. (brȧ zĭl yȧn hī-lȧndz)	4	14°00′S	48°00′W
Brazzaville, Congo (brȧ-zȧ-vēl′)	3	4°16′S	15°17′E
Brisbane, Austl. (brĭz′bān)	3	27°30′S	153°10′E
Bristol Bay, b., Ak., U.S. (brĭs′tŭl)	61	58°05′N	158°54′W
British Isles, is., Eur.	5	54°00′N	4°00′W
Brooks Range, mts., Ak., U.S. (bro͝oks)	61	68°20′N	159°00′W
Brunei, nation, Asia (brò-nī′)	3	4°52′N	113°38′E
Brussels, Bel. (brŭs′ĕls)	6	50°51′N	4°21′E
Bucharest, Rom.	6	44°23′N	26°10′E
Budapest, Hung. (bo͞o′dȧ-pĕsht′)	6	47°30′N	19°05′E
Buenos Aires, Arg. (bwā′nōs ī′rās)	2	34°20′S	58°30′W
Buffalo, N.Y., U.S. (bŭf′ȧ lō)	60	42°54′N	78°51′W
Bulgaria, nation, Eur. (bòl-gā′rĭ-ă)	3	42°12′N	24°13′E
Burbank, Ca., U.S. (bûr′bănk)	43a	34°11′N	118°19′W
Burkina Faso, nation, Afr.	68	13°00′N	2°00′W
Burma see Myanmar, nation, Asia	3	21°00′N	95°15′E
Burundi, nation, Afr.	68	3°00′S	29°30′E
Butte, Mt., U.S. (būt)	60	46°00′N	112°31′W

C

PLACE (Pronunciation)	PAGE	LAT.	LONG.
Cabrillo National Monument, rec., Ca., U.S. (kä-brēl′yō)	58	32°41′N	117°03′W
Caicos Islands, is., T./C. Is. (kī′kōs)	66	21°45′N	71°50′W
Cairo, Egypt (kī′rō)	3	30°00′N	31°17′E
Calgary, Can. (kăl′gȧ-rĭ)	60	51°03′N	114°05′W
California, Golfo de, b., Mex. (gôl-fô-dĕ-kä-lē-fôr-nyä)	61	30°30′N	113°45′W
Cambodia, nation, Asia	3	12°15′N	104°00′E
Cameroon, nation, Afr.	68	5°48′N	11°00′E
Cameroon Mountain, mtn., Cam.	5	4°12′N	9°11′E
Campbell, is., N.Z. (kăm′bĕl)	5	52°30′S	169°00′E
Campeche, Bahía de, b., Mex. (bä-ē′ä-dĕ-käm-pā′chā)	61	19°30′N	93°40′W
Canada, nation, N.A. (kăn′ȧ-dȧ)	60	50°00′N	100°00′W
Canarias, Islas (Canary Is.), is., Spain (ē′s-läs-kä-nä′ryäs)	4	29°15′N	16°30′W
Canary Islands see Canarias, Islas, is., Spain	2	29°15′N	16°30′W
Canaveral, Cape, c., Fl., U.S.	59	28°30′N	80°23′W
Canberra, Austl. (kăn′bĕr-ȧ)	3	35°21′S	149°10′E
Canyon de Chelly National Monument, rec., Az., U.S.	58	36°14′N	110°00′W
Canyonlands National Park, rec., Ut., U.S.	58	38°10′N	110°00′W
Cape Breton, i., Can. (kāp brĕt′ŭn)	60	45°48′N	59°50′W
Cape Krusenstern National Monument, rec., Ak., U.S.	58	67°30′N	163°40′W
Cape Verde, nation, Afr.	68b	15°48′N	26°02′W
Capitol Reef National Park, rec., Ut., U.S. (kăp′ĭ-tŏl)	58	38°15′N	111°10′W
Capulin Mountain National Monument, rec., N.M., U.S. (kȧ-pū′lĭn)	58	36°15′N	103°58′W
Caracas, Ven. (kä-rä′käs)	2	10°30′N	66°58′W
Caribbean Sea, sea (kăr-ĭ-bē′ăn)	66	14°30′N	75°30′W
Carlsbad Caverns National Park, rec., N.M., U.S. (kärlz′băd)	58	32°08′N	104°30′W
Caroline Islands, is., Oc.	5	8°00′N	140°00′E
Carpathians, mts., Eur. (kär-pā′thĭ-ăn)	5	49°23′N	20°14′E
Carpentaria, Gulf of, b., Austl. (kär-pĕn-târ′ĭȧ)	5	14°45′S	138°50′E
Casablanca, Mor. (kä-sä-bläŋ′kä)	2	33°32′N	7°41′W
Cascade Range, mts., N.A. (kăs-kād′)	61	42°50′N	122°20′W
Caspian Depression, depr. (kăs′pĭ-ăn)	5	47°40′N	52°35′E
Caspian Sea, sea (kăs′pĭ-ăn)	5	40°00′N	52°00′E
Castillo de San Marcos National Monument, rec., Fl., U.S. (käs-tē′lyä de-sän mär-kōs)	59	29°55′N	81°25′W
Cayman Islands, dep., N.A. (kī-män′)	66	19°30′N	80°30′W
Celebes (Sulawesi), i., Indon.	5	2°15′S	120°30′E
Celebes Sea, sea, Asia	5	3°45′N	121°52′E
Central African Republic, nation, Afr.	68	7°50′N	21°00′E
Central America, reg., N.A. (ä-mĕr′ĭ-kȧ)	60	10°45′N	87°15′W
Chad, nation, Afr. (chäd)	68	17°48′N	19°00′E
Chad, Lake, l., Afr.	5	13°55′N	13°40′E
Changchun, China (chäŋ-chòn)	7	43°55′N	125°25′E
Channel Islands, is., Ca., U.S.	58	33°30′N	119°15′W
Charleston, S.C., U.S. (chärlz′tŭn)	53	32°47′N	79°56′W
Charleston, W.V., U.S.	59	38°20′N	81°35′W
Charlotte, N.C., U.S. (shär′lŏt)	43	35°15′N	80°50′W
Chatham Islands, is., N.Z. (chăt′ăm)	4	44°00′S	178°00′W
Chelyuskin, Mys, c., Russia (chĕl-yòs′-kĭn)	5	77°45′N	104°45′E
Chengdu, China (chŭŋ-do͞o)	7	30°30′N	104°10′E
Chennai (Madras), India	3	13°08′N	80°15′E
Chesapeake Bay, b., U.S. (chĕs′ȧ-pēk)	61	38°20′N	76°15′W
Cheyenne River Indian Reservation, I.R., S.D., U.S. (shī-ĕn′)	58	45°07′N	100°46′W
Chicago, Il., U.S. (shĭ-kô-gō) (chĭ-kä′gō)	60	41°49′N	87°37′W
Chihuahua, Mex. (chê-wä′wä)	42	28°37′N	106°06′W
Chile, nation, S.A. (chē′lā)	2	35°00′S	72°00′W
Chimborazo, mtn., Ec. (chêm-bô-rä′zō)	4	1°35′S	78°45′W
China, nation, Asia (chī′nȧ)	3	36°45′N	93°00′E
Chittagong, Bngl. (chĭt-ȧ-gông′)	7	22°26′N	90°51′E
Chongqing, China (chôŋ-chyĭŋ)	3	29°38′N	107°30′E
Chonos, Archipiélago de los, is., Chile	4	44°35′N	76°15′W
Christmas Island, dep., Oc.	3	10°35′S	105°40′E
Churchill, Can. (chûrch′ĭl)	60	58°50′N	94°10′W
Cincinnati, Oh., U.S. (sĭn-sĭ-nàt′ĭ)	60	39°08′N	84°30′W
Cleveland Oh., U.S. (klēv′lănd)	60	41°30′N	81°42′W
Coast Mountains, mts., N.A. (kōst)	61	54°10′N	128°00′W
Coats Land, reg., Ant. (kōts)	4	74°00′S	30°00′W
Coco, Isla del, i., C.R. (ē′s-lä-dĕl-kô-kô)	60	5°33′N	87°02′W
Coconino, Plateau, plat., Az., U.S. (kō kō nē′nō)	58	35°45′N	112°28′W
Cocos (Keeling) Islands, is., Oc. (kō′kōs) (kē′lĭng)	3	11°50′S	90°50′E
Cod, Cape, pen., Ma., U.S.	60	41°42′N	70°15′W
Colombia, nation, S.A. (kô-lôm′bē-ä)	2	3°30′N	72°30′W
Colón, Archipiélago de (Galapagos Islands), is., Ec.	4	0°10′S	87°45′W
Colorado, r., N.A. (kŏl-ô-rä′dō)	61	36°00′N	113°30′W
Colorado National Monument, rec., Co., U.S.	58	39°00′N	108°40′W
Colorado River Indian Reservation, I.R., Az., U.S.	58	34°03′N	114°02′W
Colorado Springs, Co., U.S. (kŏl-ô-rä′dō)	43	38°49′N	104°48′W
Columbia, S.C., U.S. (kô-lŭm′bĭ-ȧ)	53	34°00′N	81°00′W
Columbia, r., N.A.	61	46°00′N	120°00′W
Columbus, Oh., U.S. (kô-lŭm′bŭs)	43	40°00′N	83°00′W
Colville Indian Reservation, I.R., Wa., U.S. (kŏl′vĭl)	58	48°15′N	119°00′W
Comorin, Cape, c., India (kō′mô-rĭn)	5	8°05′N	78°05′E
Comoros, nation, Afr.	68	12°30′S	42°45′E
Concord, N.H., U.S. (kŏŋ′kŏrd)	59	43°10′N	71°30′W
Congo, nation, Afr. (kŏn′gō)	68	3°00′S	13°48′E
Congo (Zaire), r., Afr. (kŏn′gō)	5	2°00′S	17°00′E
Congo, Democratic Republic of the (Zaire), nation, Afr.	68	1°00′S	22°15′E
Cook Islands, dep., Oc.	2	20°00′S	158°00′W
Coral Sea, sea, Oc. (kŏr′ăl)	5	13°30′S	150°00′E
Corpus Christi, Tx., U.S. (kôr′pŭs krĭstē)	58	27°48′N	97°24′W
Corsica, i., Fr. (kô′r-sē-kä)	5	42°10′N	8°55′E
Costa Rica, nation, N.A. (kŏs′tȧ rē′kȧ)	66	10°30′N	84°30′W
Côte d'Ivoire (Ivory Coast), nation, Afr.	68	7°43′N	6°30′W
Crater Lake National Park, rec., Or., U.S.	58	42°58′N	122°40′W
Craters of the Moon National Monument, rec., Id., U.S. (krā′tẽr)	58	43°28′N	113°15′W
Crete, i., Grc. (krēt)	5	35°15′N	24°30′E
Crow Creek Indian Reservation, I.R., S.D., U.S.	58	44°17′N	99°17′W
Crow Indian Reservation, I.R., Mt., U.S. (krō)	58	45°26′N	108°12′W
Crozet, Îles, is., Afr. (krô-zĕ′)	5	46°20′S	51°30′E
Cuba, nation, N.A. (kū′bȧ)	66	22°00′N	79°00′W
Cubango (Okavango), r., Afr. (ko͞o-bäŋ′gō)	5	17°10′S	18°20′E
Curitiba, Braz. (ko͞o-rē-tē′bȧ)	6	25°20′S	49°15′W
Cyprus, nation, Asia (sī′prŭs)	3	35°00′N	31°00′E
Czech Republic, nation, Eur.	3	50°00′N	15°00′E

D

PLACE (Pronunciation)	PAGE	LAT.	LONG.
Dakar, Sen. (dȧ-kär′)	2	14°40′N	17°26′W
Dalian, China (lü-dä)	7	38°54′N	121°35′E
Dallas, Tx., U.S. (dăl′lȧs)	60	32°45′N	96°48′W
Damascus, Syria	6	33°30′N	36°18′E
Damāvand, Qolleh-ye, mtn., Iran	5	36°05′N	52°05′E
Dar es Salaam, Tan. (där ĕs sȧ-läm′)	3	6°48′S	39°17′E
Darling, r., Austl.	3	31°50′S	143°20′E
Darwin, Austl. (där′wĭn)	3	12°25′S	131°00′E
Davis Strait, strt., N.A. (dā′vĭs)	61	66°00′N	60°00′W
Dawson, Can. (dô′sŭn)	60	64°04′N	139°22′W
Dayton, Oh., U.S. (dā′tŭn)	43	39°54′N	84°15′W
Death Valley National Park, rec., U.S.	58	36°34′N	117°00′W
Deccan, plat., India (dĕk′ăn)	5	19°05′N	76°40′E
Denali National Park, rec., Ak., U.S.	58a	63°48′N	153°02
Denmark Strait, strt., Eur. (dĕn′märk)	61	66°30′N	27°00
Denver, Co., U.S. (dĕn′vẽr)	60	39°44′N	104°59
Des Moines, Ia., U.S. (dē moin′)	59	41°35′N	93°37
Detroit, Mi., U.S. (dē-troit′)	60	42°22′N	83°10
Devils Lake Indian Reservation, I.R., N.D., U.S. (dĕv′'lz)	58	48°08′N	99°40
Devils Tower National Monument, rec., Wy., U.S.	58	44°38′N	105°07
Devon, Can. (dĕv′ŭn)	42	53°23′N	113°43
Dhaka, Bngl. (dä′kä) (dăk′ȧ)	6	23°45′N	90°29
Dinosaur National Monument, rec., Co., U.S. (dī′nô-sôr)	58	40°45′N	109°17
Disko, i., Grnld. (dĭs′kō)	60	70°00′N	54°00
Djibouti, nation, Afr. (jē-bo͞o-tē′)	68a	11°35′N	48°08
Dniester, r., Eur.	5	48°21′N	28°10
Dominica, nation, N.A. (dō-mĭ-nē′kȧ)	66	15°30′N	60°45
Dominican Republic, nation, N.A. (dô-mĭn′ĭ-kăn)	66	19°00′N	70°45
Donets'k, Ukr.	6	48°00′N	37°35
Drake Passage, strt. (drāk păs′ĭj)	4	57°00′S	65°00
Ducie Island, i., Pit. (dü-sē′)	4	25°30′S	126°20
Duluth, Mn., U.S. (dô-lo͞oth′)	60	46°50′N	92°07
Durban, S. Afr. (dûr′băn)	3	29°48′S	31°00
Durham, N.C., U.S. (dûr′ăm)	43	36°00′N	78°55

E

PLACE (Pronunciation)	PAGE	LAT.	LONG.
East China Sea, sea, Asia	5	30°28′N	125°52
Easter Island see Pascua, Isla de, i., Chile	4	26°50′S	109°00
East Timor, nation, Asia	3	9°00′S	125°30
Ecuador, nation, S.A. (ĕk′wȧ-dôr)	2	0°00′N	78°30
Edmonton, Can.	60	53°33′N	113°28
Effigy Mounds National Monument, rec., Ia., U.S. (ĕf′ĭ-jŭ mounds)	59	43°04′N	91°15
Egypt, nation, Afr. (ē′jĭpt)	68	26°58′N	27°01
El'brus, Gora, mtn., Russia (ĕl′bròs′)	5	43°20′N	42°25
El'brus, Mount, see El'brus, Gora, mtn., Russia	5	43°20′N	42°25
Ellesmere Island, i., Can.	60	81°00′N	80°00
El Paso, Tx., U.S. (pas′ō)	60	31°47′N	106°27
El Salvador, nation, N.A.	66	14°00′N	89°30
Enderby Land, reg., Ant. (ĕn′dĕr bĭī)	5	72°00′S	52°00
Equatorial Guinea, nation, Afr.	68	2°00′N	7°15
Erie, Lake, l., N.A. (ē′rĭ)	61	42°15′N	81°25
Eritrea, nation, Afr. (ā-rē-trā′ȧ)	68	16°15′N	38°30
Escalante, Ut., U.S. (ĕs-kȧ-lăn′tē)	58	37°50′N	111°40
Essen, Ger. (ĕs′sĕn)	6	51°26′N	6°59
Estonia, nation, Eur.	3	59°10′N	25°00
Etah, Grnld. (ē′tȧ)	60	78°20′N	72°42
Ethiopa, nation, Afr. (ē-thē-ō′pē-ȧ)	68	7°53′N	37°55
Etna, Mount, vol., Italy (ĕt′nȧ)	5	37°48′N	15°00
Europe, cont. (ū′rŭp)	64	50°00′N	15°00
Everest, Mount, mtn., Asia (ĕv′ẽr-ĕst)	5	28°00′N	86°57
Everglades National Park, rec., Fl., U.S.	59	25°39′N	80°57

F

PLACE (Pronunciation)	PAGE	LAT.	LONG.
Fairbanks, Ak., U.S. (fâr′bănks)	60	64°50′N	147°48
Falkland Islands, dep., S.A. (fôk′lȧnd)	2	50°45′S	61°00
Fargo, N.D., U.S. (fär′gō)	60	46°53′N	96°48
Faroe Islands, is., Eur.	60	62°00′N	5°45
Farvel, Kap, c., Grnld.	60	60°00′N	44°00
Fernando de Noronha, Arquipélago, is., Braz.	4	3°51′S	32°25
Fiji, nation, Oc. (fē′jē)	5	18°40′S	175°00
Finland, nation, Eur. (fĭn′lȧnd)	3	62°45′N	26°13
Flathead Indian Reservation, I.R., Mt., U.S.	58	47°30′N	114°25
Flores, i., Indon. (flō′rĕzh)	5	8°14′S	121°08
Florida, Straits of, strt., N.A.	61	24°10′N	81°00
Fond du Lac Indian Reservation, I.R., Mn., U.S.	59	46°44′N	93°04
Forel, Mont, mtn., Grnld.	61	65°50′N	37°41
Fortaleza, Braz. (fôr′tä-lā′zȧ)	2	3°35′S	38°31
Fort Apache Indian Reservation, I.R., Az., U.S. (ȧ-păch′ē)	58	34°02′N	110°27
Fort Belknap Indian Reservation, I.R., Mt., U.S.	58	48°16′N	108°38
Fort Berthold Indian Reservation, I.R., N.D., U.S. (bẽrth′ôld)	58	47°47′N	103°28
Fort Frederica National Monument, rec., Ga., U.S. (frĕd′ĕ-rĭ-kȧ)	59	31°13′N	85°25
Fort Lauderdale, Fl., U.S. (lô′dẽr-dāl)	43a	26°07′N	80°09
Fort Matanzas, Fl., U.S. (mä-tän′zäs)	59	29°39′N	81°17
Fort Myers, Fl., U.S. (mī′ẽrz)	43a	26°36′N	81°45
Fort Peck Indian Reservation, I.R., Mt., U.S.	58	48°22′N	105°40
Fort Pulaski National Monument, rec., Ga., U.S. (pu-lăs′kĭ)	59	31°59′N	80°56
Fort Simpson, Can. (sĭmp′sŭn)	60	61°52′N	121°48
Fort Sumter National Monument, rec., S.C., U.S. (sŭm′tẽr)	59	32°43′N	79°54
Fort Union National Monument, rec., N.M., U.S. (ūn′yŭn)	58	35°51′N	104°57

ăt; finăl; rāte; senāte; ärm; ȧsk; sofȧ; fâre; ch-choose; dh-as th in other; bē; ēvent; bĕt; recĕnt; cratẽr; g-gō; gh-guttural g; bĭt; ĭ-short neutral; rīde; κ-guttural k as ch in German ich;

LACE (Pronunciation)	PAGE	LAT.	LONG.
ort Wayne, In., U.S. (wān)	53	41°00′N	85°10′W
ort Worth, Tx., U.S. (wûrth)	60	32°45′N	97°20′W
oxe Basin, b., Can. (fŏks)	61	67°35′N	79°21′W
rance, nation, Eur. (fråns)	3	46°39′N	0°47′E
ranz Josef Land *see* Zemlya Frantsa-Iosifa, is., Russia	5	81°32′N	40°00′E
rench Guiana, dep., S.A. (gē-ä′nä)	2	4°20′N	53°00′W
rench Polynesia, dep., Oc.	2	15°00′S	140°00′W
resno, Ca., U.S. (frĕs′nō)	53	36°44′N	119°46′W
rio, Cabo, c., Braz. (kä′bō-frē′ō)	4	22°58′S	42°08′W
uji San, mtn., Japan (fōō′jē sän)	5	35°23′N	138°44′E
ukui, Japan (fōō′kōō-ē)	7	36°05′N	136°14′E
ukuoka, Japan (fōō′kō-ō′kå)	7	33°35′N	130°23′E
undy, Bay of, b., Can. (fŭn′dĭ)	61	45°00′N	66°00′W
undy National Park, rec., Can.	59	45°38′N	65°00′W

G

abon, nation, Afr. (gå-bôɴ′)	68	0°30′S	10°45′E
alapagos Islands *see* Colón, Archipiélago de, is., Ec.	2	0°10′S	87°45′W
allinas, Punta de, c., Col. (gä-lyē′näs)	60	12°10′N	72°10′W
alveston I, Tx., U.S. (găl′vĕs-tŭn)	60	29°12′N	94°53′W
ambie, r., Afr.	2	12°30′N	13°00′W
anges, r., Asia (găn′jēz)	5	24°00′N	89°30′E
ary, In., U.S. (gā′rĭ)	53	41°35′N	87°21′W
ates of the Arctic National Park, rec., Ak., U.S.	58	67°45′N	153°30′W
eorgetown, Guy. (jôrj′toun)	2	7°45′N	58°04′W
eorge Washington Carver National Monument, rec., Mo., U.S. (jôrg wăsh-ĭng-tŭn kär′vēr)	59	36°58′N	94°21′W
eorgia, nation, Asia (jôr′ji-ă)	3	42°17′N	43°00′E
eorgian Bay Islands National Park, rec., Can.	59	45°20′N	81°40′W
ermany, nation, Eur. (jûr′må-nĭ)	3	51°00′N	10°00′E
hana, nation, Afr. (gän′ä)	68	8°00′N	2°00′W
ibraltar, Strait of, strt. (gĭ-bräl-tä′r)	4	35°55′N	5°45′W
ila Cliff Dwellings National Monument, rec., N.M., U.S. (hē′lå)	58	33°15′N	108°20′W
ila River Indian Reservation, I.R., Az., U.S.	58	33°11′N	112°38′W
ilbert Islands, is., Kir. (gĭl-bĕrt)	5	0°30′S	174°00′E
lacier Bay National Park, rec., Ak., U.S. (glā′shēr)	58a	58°40′N	136°50′W
lacier National Park, rec., Can.	58	51°45′N	117°35′W
len Canyon National Recreation Area, rec., U.S.	58	37°00′N	111°20′W
obi , des., Asia (gō′be)	5	43°29′N	103°15′E
odhavn, Grnld. (gōdh′håvn)	60	69°15′N	53°30′W
odthåb, Grnld. (gŏt′hōb)	60	64°10′N	51°32′W
ood Hope, Cape of, c., S. Afr. (kāp ov gŏŏd hōp)	5	34°21′S	18°29′E
ough, i., St. Hel. (gŏf)	4	40°00′S	10°00′W
rand Canyon National Park, rec., Az., U.S.	58	36°15′N	112°20′W
rand Canyon-Parashant National Monument, rec., Az., U.S.	58	36°25′N	113°45′W
rand Portage Indian Reservation, I.R., Mn., U.S. (pōr′tĭj)	59	47°54′N	89°34′W
rand Rapids, Mi., U.S. (răp′ĭdz)	53	43°00′N	85°45′W
rand Staircase-Escalante National Monument, rec., Ut., U.S.	58	37°25′N	111°30′W
rand Teton National Park, rec., Wy., U.S. (tē′tŏn)	58	43°54′N	110°15′W
reat Australian Bight, b., Austl. (ôs-trā′lĭ-ăn bĭt)	5	33°30′S	127°00′E
reat Barrier Reef, rf., Austl. (bå-rĭ-ĕr rēf)	5	16°43′S	146°34′E
reat Basin, basin, U.S. (grāt bā′s′n)	61	40°08′N	117°10′W
reat Bear Lake, l., Can. (bâr)	61	66°10′N	119°53′W
reat Dividing Range, mts., Austl. (dĭ-vī-dĭng rānj)	5	35°16′S	146°38′E
reater Antilles, is., N.A.	4	20°30′N	79°15′W
reater Khingan Range, mts., China (dä hĭŋ-gän lĭŋ)	5	46°30′N	120°00′E
reat Indian Desert, des., Asia	5	27°35′N	71°37′E
reat Karroo, plat., S. Afr. (grāt kå′rōō)	5	32°45′S	22°00′E
reat Plains, pl., N.A. (plāns)	61	45°00′N	104°00′W
reat Salt Lake, l., Ut., U.S. (sôlt lāk)	61	41°19′N	112°48′W
reat Sandy Desert, des., Austl. (săn′dē)	5	21°50′S	123°10′E
reat Slave Lake, l., Can. (slāv)	61	61°37′N	114°58′W
reat Smoky Mountains National Park, rec., U.S. (smōk-ē)	59	35°43′N	83°20′W
reat Victoria Desert, des., Austl. (vĭk-tō′rĭ-å)	5	29°45′S	124°30′E
reece, nation, Eur. (grēs)	3	39°00′N	21°30′E
reenland, dep., N.A. (grēn′lănd)	60	74°00′N	40°00′W
reenland Sea, sea	61	77°00′N	1°00′W
reen Mountains, mts., N.A.	59	43°10′N	73°05′W
reensboro, N.C., U.S.	43	36°04′N	79°45′W
reenville, S.C., U.S.	53	34°50′N	82°25′W
renada, nation, N.A. (grē-nä′da)	66	12°02′N	61°15′W
renadines, The, is., N.A. (grĕn′å-dēnz)	66b	12°37′N	61°35′W
uadalajara, Mex. (gwä-dhä-lä-hä′rä)	60	20°41′N	103°21′W
uadalupe, i., Mex.	60	29°00′N	118°45′W
uadalupe Mountains, mts., N.M., U.S.	58	32°00′N	104°55′W
uadeloupe, dep., N.A. (gwä-dē-lōōp)	66	16°40′N	61°10′W
uam, i., Oc. (gwäm)	3	14°00′N	143°20′E
Guangzhou (Canton), China	3	23°07′N	113°15′W
Guatemala, Guat. (guä-tå-mä′lä)	66	14°37′N	90°32′W
Guatemala, nation, N.A.	66	15°45′N	91°45′W
Guayaquil, Golfo de, b., Ec. (gōl-fô-dĕ)	4	3°03′S	82°12′W
Guiana Highlands, mts., S.A.	4	3°20′N	60°00′W
Guinea, nation, Afr. (gĭn′ē)	68	10°48′N	12°28′W
Guinea, Gulf of, b., Afr.	5	2°00′N	1°00′E
Guinea-Bissau, nation, Afr. (gĭn′ē)	68	12°00′N	20°00′W
Gunnison, Ut., U.S. (gŭn′ĭ-sŭn)	58	39°10′N	111°50′W
Guyana, nation, S.A. (gŭy′ănă)	2	7°45′N	59°00′W
Gwardafuy, Gees, c., Som.	5	11°55′N	51°30′E

H

Ḩadd, Ra's al, c., Oman	5	22°29′N	59°46′E
Hainan Dao, i., China (hī-nän dou)	5	19°00′N	111°10′E
Haiti, nation, N.A. (hā′tĭ)	66	19°00′N	72°15′W
Halifax, Can. (hăl′ĭ-făks)	60	44°39′N	63°36′W
Halmahera, i., Indon. (häl-mä-hā′rä)	5	1°00′S	129°00′E
Hamburg, Ger. (häm′bōōrgh)	6	53°34′N	10°02′E
Hamilton, Can. (hăm′ĭl-tŭn)	53	43°15′N	79°52′W
Hanoi, Viet. (hä-noi′)	3	21°04′N	105°50′E
Harbin, China	3	45°40′N	126°30′E
Harrisburg, Pa., U.S.	53	40°15′N	76°50′W
Hartford, Ct., U.S. (härt′fẽrd)	43	41°45′N	72°40′W
Hatteras, Cape, c., N.C., U.S. (hăt′ēr-ås)	60	35°15′N	75°24′W
Havana, Cuba	60	23°08′N	82°23′W
Hawaii, state, U.S.	2	20°00′N	157°40′W
Heard Island, i., Austl. (hûrd)	5	53°10′S	74°35′E
Hekla, vol., Ice.	61	63°53′N	19°37′W
Hermosillo, Mex. (ĕr-mô-sē′l-yŏ)	66	29°00′N	110°57′W
Himalayas, mts., Asia	5	29°30′N	85°02′E
Hindu Kush, mts., Asia (hĭn′dōō kōōsh′)	5	35°15′N	68°44′E
Hispaniola, i., N.A. (hĭ′spăn-ĭ-ō-lå)	60	17°30′N	73°15′W
Ho Chi Minh City, Viet.	3	10°46′N	106°34′E
Hokkaidō, i., Japan (hōk′kī-dō)	5	43°30′N	142°45′E
Honduras, nation, N.A. (hŏn-dōō′rås)	66	14°30′N	88°00′W
Honduras, Gulf of, b., N.A.	61	16°30′N	87°30′W
Hong Kong (Xianggang), China	3	21°45′N	115°00′E
Honolulu, Hi., U.S. (hŏn-ō-lōō′lōō)	2	21°18′N	157°50′W
Honshū, i., Japan	5	36°00′N	138°00′E
Hood, Mount, mtn., Or., U.S.	58	45°20′N	121°43′W
Horn, Cape *see* Hornos, Cabo de, c., Chile	4	56°00′S	67°00′W
Hornos, Cabo de (Cape Horn), c., Chile	4	56°00′S	67°00′W
Hot Springs National Park, rec., Ar., U.S.	59	34°30′N	93°00′W
Houston, Tx., U.S. (hūs′tŭn)	60	29°46′N	95°21′W
Howe, Cape, c., Austl. (hou)	5	37°30′S	150°40′E
Howland, i., Oc. (hou′lănd)	2	1°00′N	176°00′W
Hualapai Indian Reservation, I.R., Az., U.S. (wälåpī)	58	35°41′N	113°38′W
Huang (Yellow), r., China (hůäŋ)	5	35°06′N	113°39′E
Hudson Bay, Can. (hŭd′sŭn)	61	52°52′N	102°25′W
Hudson Bay, b., Can.	61	60°15′N	85°30′W
Hungary, nation, Eur. (hŭŋ′gå-rĭ)	3	46°44′N	17°55′E
Huron, Lake, l., N.A. (hū′rŏn)	61	45°15′N	82°40′W
Hyderābād, India (hī-dēr-å-băd′)	3	17°29′N	78°28′E

I

Iceland, nation, Eur. (īs′lănd)	60	65°12′N	19°45′W
Iguidi, Erg, Afr.	4	26°22′N	6°53′W
India, nation, Asia (ĭn′dĭ-å)	3	23°00′N	77°30′E
Indianapolis, In., U.S. (ĭn-dĭ-ăn-ăp′ō-lĭs)	43	39°45′N	86°08′W
Indian Ocean, o.	5	10°00′S	70°00′E
Indonesia, nation, Asia (ĭn′dō-nē-zhå)	3	4°38′S	118°45′E
Indus, r., Asia (ĭn′dŭs)	5	26°43′N	67°41′E
Inuvik, Can.	60	68°40′N	134°10′W
Iqaluit, Can.	60	63°48′N	68°31′W
Iran, nation, Asia (ē-rän′)	3	31°15′N	53°30′E
Iran, Plateau of, plat., Iran	5	32°28′N	58°00′E
Iraq, nation, Asia (ē-räk′)	3	32°00′N	42°30′E
Irazú, vol., C.R. (ē-rä-zōō′)	4	9°58′N	83°54′W
Ireland, nation, Eur. (īr-lănd)	3	53°33′N	8°00′W
Irkutsk, Russia (ĭr-kŏŏtsk′)	3	52°16′N	104°00′E
Irtysh, r., Asia (ĭr-tĭsh′)	5	59°00′N	69°00′E
Isabella Indian Reservation, I.R., Mi., U.S. (ĭs-å-bĕl′-lå)	59	43°35′N	84°55′W
Isle Royale National Park, rec., Mi., U.S. (īl′roi-ăl′)	59	47°57′N	88°37′W
Israel, nation, Asia	3	32°40′N	34°00′E
İstanbul, Tur. (ē-stän-bōōl′)	6	41°02′N	29°00′E
Italy, nation, Eur. (ĭt′å-lē)	3	43°58′N	11°14′E
Ivory Coast *see* Cote d'Ivoire, nation, Afr.	68	7°43′N	6°30′W
Ivvavik National Park, rec., Can.	58	69°10′N	139°30′W

J

Jackson, Ms., U.S. (jăk′sŭn)	59	32°17′N	90°10′W
Jacksonville, Fl., U.S. (jăk′sŭn-vĭl)	60	30°20′N	81°40′W
Jaipur, India	7	27°00′N	75°50′E
Jakarta, Indon. (yä-kär′tä)	3	6°17′S	106°45′E
Jamaica, nation, N.A.	66	17°45′N	78°00′W
James Bay, b., Can. (jāmz)	61	53°53′N	80°40′W
Jan Mayen, i., Nor. (yän mī′ĕn)	60	70°59′N	8°05′W
Japan, nation, Asia (jå-păn′)	3	36°30′N	133°30′E
Japan, Sea of, sea, Asia (jå-păn′)	5	40°08′N	132°55′E
Java Trench, deep	5	9°45′S	107°30′E
Jawa, Laut (Java Sea), sea, Indon.	5	5°10′S	110°30′E
Jicarilla Apache Indian Reservation, I.R., N.M., U.S. (ᴋē-kå-rēl′yå)	58	36°45′N	107°00′W
Jinan, China (jyē-nän)	7	36°40′N	117°01′E
Johannesburg, S. Afr. (yō-hän′ĕs-bōrgh)	3	26°08′S	27°54′E
Johnston, i., Oc. (jŏn′stŭn)	2	17°00′N	168°00′W
Jordan, nation, Asia (jôr′dăn)	3	30°15′N	38°00′E
Joshua Tree National Park, rec., Ca., U.S. (jō′shū-å trē)	58	34°02′N	115°53′W
Juan Fernández, Islas de, is., Chile	4	33°30′S	79°00′W
Julianehåb, Grnld.	60	60°07′N	46°20′W
Juneau, Ak., U.S. (jōō′nō)	60	58°25′N	134°30′W

K

Kābul, Afg. (kä′bŏŏl)	7	34°39′N	69°14′E
Kalahari Desert, des., Afr. (kä-lä-hä′rē)	5	23°00′S	22°03′E
Kamchatka, Poluostrov, pen., Russia	5	55°19′N	157°45′E
Kānpur, India (kän′pûr)	7	26°30′N	80°10′E
Kansas City, Mo., U.S. (kăn′zås)	60	39°05′N	94°35′W
Karāchi, Pak.	3	24°59′N	68°56′E
Karskoye More (Kara Sea), sea, Russia	5	74°00′N	68°00′E
Kasai (Cassai), r., Afr.	5	3°45′S	19°10′E
Katmai National Park, rec., Ak., U.S. (kăt′mī)	58a	58°38′N	155°00′W
Katowice, Pol.	6	50°15′N	19°00′E
Kazakhstan, nation, Asia	3	48°45′N	59°00′E
Kenai Fjords National Park, rec., Ak., U.S. (kĕ-nī′)	58	59°45′N	150°00′W
Kenya, nation, Afr. (kēn′yå)	68	1°00′N	36°53′E
Kerguélen, Îles, is., Afr. (kĕr′gå-lĕn)	5	49°50′S	69°30′E
Kermadec Islands, is., N.Z. (kẽr-måd′ĕk)	4	30°30′S	177°00′E
Kiev (Kyïv), Ukr.	3	50°27′N	30°30′E
Kilimanjaro, mtn., Tan. (kyl-ē-măn-jä′rō)	5	3°09′S	37°19′E
Kinabalu, Gunong, mtn., Malay.	5	5°45′N	115°26′E
Kings Canyon National Park, rec., Ca., U.S. (kăn′yŭn)	58	36°52′N	118°53′W
Kingston, Jam. (kĭngz′tŭn)	60	18°00′N	76°45′W
Kinshasa, D.R.C.	3	4°18′S	15°18′E
Kiribati, nation, Oc.	3	1°30′S	173°00′E
Kiritimati, i., Kir.	4	2°20′N	157°40′W
Klondike Region, hist. reg., N.A. (klŏn′dīk)	60	64°12′N	142°38′W
Kluane National Park, rec., Can.	58	60°25′N	137°53′W
Klyuchevskaya, vol., Russia (klyōō-chĕfskä′yä)	5	56°13′N	160°00′E
Knoxville, Tn., U.S. (nŏks′vĭl)	53	35°58′N	83°55′W
Kobuk Valley National Park, rec., Ak., U.S. (kō′bŭk)	58	67°20′N	159°00′W
Kodiak Island, i., Ak., U.S. (kō′dyăk)	60	57°24′N	153°32′W
Kolkata (Calcutta), India	3	22°32′N	88°22′E
Kootenay National Park, rec., Can. (kōō′tē-nå)	58	51°06′N	117°02′W
Korea, North, nation, Asia	3	40°00′N	127°00′E
Korea, South, nation, Asia	3	36°30′N	128°00′E
Kosciuszko, Mount, mtn., Austl. (kŏs-ĭ-ŭs′kō)	5	36°26′S	148°20′E
Kouchibouguac National Park, rec., Can.	59	46°53′N	65°35′W
Kra, Isthmus of, isth., Asia	5	9°30′S	99°45′E
Kuala Lumpur, Malay. (kwä′lä lŏŏm-pōōr′)	3	3°08′N	101°42′E
Kunlun Shan, mts., China (kōōn-lōōn shän)	5	35°26′N	83°09′E
Kuril Islands, is., Russia (kōō′rĭl)	5	46°20′N	149°30′E
Kutch, Gulf of, b., India	5	22°45′N	68°33′E
Kuwait, nation, Asia	3	29°00′N	48°45′E
Kyrgyzstan, nation, Asia	3	41°45′N	74°38′E
Kyūshū, i., Japan	5	33°00′N	131°00′E

L

Labrador, reg., Can. (lăb′rå-dôr)	60	53°05′N	63°30′W
Labrador Sea, sea, Can.	42	50°38′N	55°00′W
Lac Court Oreille Indian Reservation, I.R., Wi., U.S.	59	46°04′N	91°18′W
Lac du Flambeau Indian Reservation, I.R., Wi., U.S.	59	46°12′N	89°50′W
Ladozhskoye Ozero, l., Russia (lå-dôsh′skô-yē ô′zĕ-rô)	5	60°59′N	31°30′E
Lagos, Nig. (lä′gōs)	3	6°27′N	3°24′E

g-sing; ŋ-baŋk; ɴ-nasalized n; nŏd; cŏmmit; ōld; ōbey; ôrder; oi-boil; fōōd; ŏŏ-as oo in foot; ou-out; s-soft; sh-dish; th-thin; pūre; ůnite; ûrn; stŭd; circŭs; ü-as in French tu; ′-indeterminate vowel.

PLACE (Pronunciation)	PAGE	LAT.	LONG.
Lahore, Pak. (lä-hōr′)	7	32°00′N	74°18′E
Lake Clark National Park, rec., Ak., U.S.	58	60°30′N	153°15′W
Lake Mead National Recreation Area, rec., U.S.	58	36°00′N	114°30′W
Lakshadweep, is., India	5	11°00′N	73°02′E
Lands End, c., Eng., U.K.	5	50°03′N	5°45′W
L'Anse and Vieux Desert Indian Reservation, I.R., Mi., U.S. (làns)	59	46°41′N	88°12′W
Laos, nation, Asia (lā-ōs) (là-ōs′)	3	20°15′N	102°00′E
La Paz, Bol. (lä päz′)	2	16°31′S	68°03′W
La Paz, Mex. (lä-pà′z)	42	24°00′N	110°15′W
Lassen Volcanic National Park, rec., Ca., U.S. (lăs′ĕn)	58	40°43′N	121°35′W
Las Vegas, Nv., U.S. (läs vā′gäs)	43	36°12′N	115°10′W
Latvia, nation, Eur.	3	57°28′N	24°29′E
Laurentian Highlands, hills, Can. (lô′rĕn-tĭ-ăn)	61	49°00′N	74°50′W
Lebanon, nation, Asia (lĕb′*à*-nŭn)	3	34°00′N	34°00′E
Leech, l., Mn., U.S. (lēch)	59	47°06′N	94°16′W
Leeuwin, Cape, c., Austl. (lōō′wĭn)	5	34°15′S	114°30′E
Lena, r., Russia	5	68°00′N	123°00′E
Lesotho, nation, Afr. (lĕsō′thŏ)	68	29°45′S	28°07′E
Lesser Antilles, is.	4	12°15′N	65°00′W
Liberia, nation, Afr. (lī-bē′rĭ-*à*)	68	6°30′N	9°55′W
Libya, nation, Afr. (lĭb′ē-ä)	68	27°38′N	15°00′E
Libyan Desert, des., Afr. (lĭb′ē-*ă*n)	5	28°23′N	23°34′E
Lima, Peru (lē′mä)	2	12°06′S	76°55′W
Lisbon (Lisboa), Port.	2	38°42′N	9°05′W
Lithuania, nation, Eur. (lĭth-ů-ā′nĭ-*à*)	3	55°42′N	23°30′E
Little Missouri, r., U.S. (mĭ-sōō′rĭ)	58	46°00′N	104°00′W
Little Rock, Ar., U.S. (rôk)	43	34°42′N	92°16′W
Logan, Mount, mtn., Can. (lō′g*ă*n)	61	60°54′N	140°33′W
London, Can. (lŭn′d*ŭ*n)	59	43°00′N	81°20′W
London, Eng., U.K.	3	51°30′N	0°07′W
Lookout Point Lake, res., Or., U.S. (lòkôut)	59	43°51′N	122°38′W
Los Angeles, Ca., U.S. (äŋ′hå-lās)	60	34°03′N	118°14′W
Louisville, Ky., U.S. (lōō′ĭs-vĭl) (lōō′ē-vĭl)	43	38°15′N	85°45′W
Lower Brule Indian Reservation, I.R., S.D., U.S. (brü′lā)	58	44°15′N	100°21′W
Luanda, Ang. (lōō-än′dä)	3	8°48′S	13°14′E
Lubbock, Tx., U.S.	58	33°35′N	101°50′E
Lucknow, India (lŭk′nou)	7	26°54′N	80°58′E
Luzon, i., Phil. (lōō-zŏn′)	5	17°10′N	119°45′E

M

PLACE (Pronunciation)	PAGE	LAT.	LONG.
Mackenzie, r., Can. (m*à*-kĕn′zĭ)	61	63°38′N	124°23′W
Macquarie Islands, is., Austl. (m*à*-kwŏr′ē)	5	54°36′S	158°45′E
Madagascar, nation, Afr. (măd-*à*-găs′k*à*r)	68	18°05′S	43°12′E
Madeira, Arquipélago da, is., Port.	2	33°26′N	16°44′W
Madeira, Ilha da, i., Port. (mä-dā′rä)	4	32°41′N	16°15′W
Madre, Sierra, mts., N.A. (sē-ĕ′r-rä-mà′drĕ)	61	15°55′N	92°40′W
Madre del Sur, Sierra, mts., Mex. (sē-ĕ′r-rä-mä′drå dĕlsōōr′)	42	17°35′N	100°35′W
Madrid, Spain (mä-drē′d)	3	40°26′N	3°42′W
Magadan Oblast, Russia (må-gå-dän′)	3	65°00′N	160°00′E
Magallanes, Estrecho de, strt., S.A.	4	52°30′S	68°45′W
Magdalena, r., Col.	61	7°45′N	74°04′W
Makah Indian Reservation, I.R., Wa., U.S.	58	48°17′N	124°52′W
Malacca, Strait of, strt., Asia (m*à*-lăk′*à*)	5	4°15′N	99°44′E
Malawi, nation, Afr.	68	11°15′S	33°45′E
Malay Peninsula, pen., Asia (m*à*-lā′) (mā′lā)	3	6°00′N	101°00′E
Malaysia, nation, Asia (m*à*-lā′zh*à*)	3	4°10′N	101°22′E
Malden, i., Kir. (môl′dĕn)	4	4°20′S	154°30′W
Maldives, nation, Asia	3	4°30′N	71°30′E
Mali, nation, Afr.	68	15°45′N	0°15′W
Malpelo, Isla de, i., Col. (mäl-pā′lō)	60	3°55′N	81°30′W
Maluku (Moluccas), is., Indon.	5	2°22′S	128°25′E
Mammoth Cave National Park, rec., Ky., U.S. (măm′*ŏ*th)	59	37°20′N	86°21′W
Managua, Nic. (mä-nä′gwä)	66	12°10′N	86°16′W
Manaus, Braz. (mä-nä′ōōzh)	2	3°01′S	60°00′W
Manchester, Eng., U.K. (măn′chĕs-tẽr)	6	53°28′N	2°14′W
Manchester, Ma., U.S.	43a	42°35′N	70°47′W
Manihiki Islands, is., Cook Is. (mä′nē-hē′kē)	4	9°40′S	158°00′W
Manila, Phil. (m*à*-nĭl′*à*)	3	14°37′N	121°00′E
Maputo, Moz.	3	26°50′S	32°30′E
Maracaibo, Lago de, l., Ven. (lä′gô-dĕ-mä-rä-kī′bō)	4	9°55′N	72°13′W
Marajó, Ilha de, i., Braz.	4	1°00′S	49°30′W
Mariana Islands, is., Oc. (mä-ryä′nä)	3	16°00′N	145°30′E
Mariana Trench, deep	5	12°00′N	144°00′E
Marquesas Islands, is., Fr. Poly. (mär-kē′säs)	4	8°50′S	141°00′W
Marshall Islands, nation, Oc. (mär′sh*ă*l)	3	10°00′N	165°00′E
Martinique, dep., N.A. (mår-tē-nēk′)	66	14°50′N	60°40′W
Mascarene Islands, is., Afr.	5	20°20′S	56°40′E
Mato Grosso, Chapada de, hills, Braz. (shä-pä′dä-dĕ mät′ō grōs′ō)	4	13°39′S	55°42′W
Mauna Kea, mtn., Hi., U.S. (mä′ō-näkä′ä)	4	19°52′N	155°30′W
Mauricie, Parc National de la, rec., Can.	59	46°46′N	73°00′W
Mauritania, nation, Afr. (mô-rē-tā′nĭ-*à*)	68	19°38′N	13°30′W
Mauritius, nation, Afr. (mô-rĭsh′ĭ-*ŭ*s)	68	20°18′S	57°36′E
Mazatlán, Mex. (mä-zä-tlän′)	42	23°14′N	106°27′W
McAllen, Tx., U.S. (m*ă*k-ăl′ĕn)	53	26°12′N	98°14′W
McKinley, Mount, mtn., Ak., U.S. (m*à*-kĭn′lĭ)	61	63°00′N	151°02′W
Mecca (Makkah), Sau. Ar. (mĕk′*à*)	3	21°27′N	39°45′E
Medellín, Col. (må-dhĕl-yēn′)	6	6°15′N	75°34′W
Mediterranean Sea, sea (mĕd-ĭ-tẽr-ā′nē-*ă*n)	5	36°22′N	13°25′E
Melbourne, Austl. (mĕl′b*ŭ*rn)	3	37°52′S	145°08′E
Melville, l., Can. (mĕl′vĭl)	42	53°46′N	59°31′W
Memphis, Tn., U.S. (mĕm′fĭs)	60	35°07′N	90°03′W
Mendocino, Cape, c., Ca., U.S. (mĕn′dō-sē′nō)	60	40°25′N	12°42′W
Mérida, Mex.	66	20°58′N	89°37′W
Mexico, nation, N.A.	66	23°45′N	104°00′W
Mexico, Gulf of, b., N.A.	61	25°15′N	93°45′W
Mexico City, Mex. (mĕk′sĭ-kō)	60	19°28′N	99°09′W
Miami, Fl., U.S.	60	25°45′N	80°11′W
Michigan, Lake, l., U.S.	61	43°20′N	87°10′W
Micronesia, Federated States of, nation, Oc.	3	5°00′N	152°00′E
Midway Islands, is., Oc.	2	28°00′N	179°00′W
Milan (Milano), Italy (mė-lä′nō)	6	45°29′N	9°12′E
Milwaukee, Wi., U.S.	60	43°03′N	87°55′W
Mindanao, i., Phil.	5	8°00′N	125°00′E
Minneapolis, Mn., U.S. (mĭn-ē-ăp′ô-lĭs)	60	44°58′N	93°15′W
Mississippi, r., U.S. (mĭs-ĭ-sĭp′ē)	61	32°00′N	91°30′W
Missouri, r., U.S. (mĭ-sōō′rē)	61	40°40′N	96°00′W
Mitchell, Mount, mtn., N.C., U.S. (mĭch′*ĕ*l)	4	35°47′N	82°15′W
Mobile, Al., U.S. (mō-bēl′)	60	30°42′N	88°03′W
Mogadishu (Muqdisho), Som.	3	2°08′N	45°22′E
Mojave Desert, des., Ca., U.S. (mō-hä′vä)	58	35°00′N	117°00′W
Moldova, nation, Eur.	3	48°00′N	28°00′E
Moluccas *see* Maluku, is., Indon.	5	2°22′S	128°25′E
Mombasa, Kenya (mōm-bä′sä)	3	4°03′S	39°40′E
Mongolia, nation, Asia (mŏŋ-gō′lĭ-*à*)	3	46°00′N	100°00′E
Monterrey, Mex. (mōn-tĕr-rā′)	66	25°43′N	100°19′W
Montevideo, Ur. (mōn′tå-vē-dhā′ō)	2	34°50′S	56°10′W
Montréal, Can. (mŏn-trē-ôl′)	60	45°30′N	73°35′W
Montserrat, dep., N.A. (mŏnt-sĕ-răt′)	66	16°48′N	63°15′W
Morocco, nation, Afr. (mō-rŏk′ō)	68	32°00′N	7°00′W
Moscow (Moskva), Russia	3	55°45′N	37°37′E
Mount Rainier National Park, rec., Wa., U.S. (rå-nēr′)	58	46°47′N	121°17′W
Mozambique, nation, Afr. (mō-z*ă*m-bēk′)	68	20°15′S	33°53′E
Mozambique Channel, strt., Afr. (mō-z*ă*m-bek′)	5	24°00′S	38°00′E
Mumbai (Bombay), India	3	18°58′N	72°50′E
Myanmar (Burma), nation, Asia	3	21°00′N	95°15′E

N

PLACE (Pronunciation)	PAGE	LAT.	LONG.
Nagoya, Japan	7	35°09′N	136°53′E
Nāgpur, India (näg′pōōr)	7	21°12′N	79°09′E
Nahanni National Park, rec., Can.	58	62°10′N	125°15′W
Nairobi, Kenya (nī-rō′bē)	3	1°17′S	36°49′E
Namibia, nation, Afr.	68	19°30′S	16°13′E
Nanjing, China (nän-jyĭŋ)	3	32°04′N	118°46′E
Nansei-shotō, is., Japan	5	27°30′N	127°00′E
Nashville, Tn., U.S. (năsh′vĭl)	42	36°10′N	86°48′W
Nassau, Bah. (năs′ô)	66	25°05′N	77°20′W
Nauru, nation, Oc.	3	0°30′S	167°00′E
Navajo Hopi Joint Use Area, I.R., Az., U.S.	58	36°15′N	110°30′W
Navajo Indian Reservation, I.R., U.S. (năv′*à*-hō)	58	36°31′N	109°24′W
Navajo National Monument, rec., Az., U.S.	58	36°43′N	110°39′W
Negro, r., S.A. (nā′grō)	61	0°18′S	63°21′W
Nelson, Can. (nĕl′s*ŭ*n)	60	49°29′N	117°17′W
Nelson, r., Can.	61	56°50′N	93°40′W
Nepal, nation, Asia (nĕ-pôl′)	3	28°45′N	83°00′E
Netherlands, nation, Eur. (nĕdh′ẽr-l*ă*ndz)	3	53°01′N	3°57′E
Nettilling, l., Can.	61	66°30′N	70°40′W
Nevada, Sierra, mts., U.S. (sē-ĕ′rä nē-vä′d*à*)	61	39°20′N	120°05′W
Nevis, i., St. K./N. (nē′vĭs)	66	17°05′N	62°38′W
Newark, N.J., U.S. (nōō′ûrk)	6	40°44′N	74°10′W
New Britain, i., Pap. N. Gui. (brĭt′′n)	5	6°45′S	149°38′E
New Caledonia, dep., Oc.	3	21°28′S	164°40′E
New Delhi, India (dĕl′hĭ)	3	28°43′N	77°18′E
Newfoundland, i., Can.	60	48°30′N	56°00′W
New Guinea, i., (gĭne)	3	5°45′S	140°00′E
New Haven, Ct., U.S. (hā′vĕn)	53	41°20′N	72°55′W
New Hebrides, is., Vanuatu	5	16°00′S	167°00′E
New Ireland, i., Pap. N. Gui. (īr′l*ă*nd)	5	3°15′S	152°30′E
New Orleans, La., U.S. (ôr′lē-*ă*nz)	60	30°00′N	90°05′W
New York, N.Y., U.S. (yôrk)	60	40°40′N	73°58′W
New Zealand, nation, Oc. (zē′lănd)	3	42°00′S	175°00′E
Nez Perce Indian Reservation, I.R., Id., U.S.	58	46°20′N	116°30′W
Nicaragua, nation, N.A. (nĭk-*à*-rä′gw*à*)	66	12°45′N	86°15′W
Nicaragua, Lago de, l., Nic. (lä′gô dĕ)	61	11°45′N	85°28′W
Nicobar Islands, is., India (nĭk-ô-bär′)	5	8°28′N	94°04
Niger, nation, Afr. (nī′jẽr)	68	18°02′N	8°30
Niger, r., Afr.	5	8°00′N	6°00
Nigeria, nation, Afr. (nī-jē′rĭ-*à*)	68	8°57′N	6°30
Nile, r., Afr. (nīl)	5	27°30′N	31°00
Nipigon, l., Can. (nĭp′ĭ-gŏn)	61	49°37′N	89°55′
Nome, Ak., U.S. (nōm)	60	64°30′N	165°20
Noranda, Can.	59	48°15′N	79°01
Nord Kapp, c., Nor.	5	71°11′N	25°48
Norfolk, Va., U.S. (nôr′f*ŏ*k)	60	36°55′N	76°15
North America, cont.	61	45°00′N	100°00
North American Basin, deep, (*à*-mĕr′ĭ-k*ă*n)	4	23°45′N	62°45′
North Cape, c., N.Z.	5	34°31′S	173°02
North Cascades National Park, rec., Wa., U.S.	58	48°50′N	120°50′
Northern Cheyenne Indian Reservation, I.R., Mt., U.S.	58	45°32′N	106°43
North Island, i., N.Z.	5	37°20′S	173°30
North Magnetic Pole, pt. of. i.	60	77°19′N	101°49
North Pole, pt. of. i.	61	90°00′N	0°00′
North Sea, Eur.	5	56°09′N	3°16′
North West Cape, c., Austl. (nôrth′wĕst)	5	21°50′S	112°25′
Norway, nation, Eur. (nôr′wā)	3	63°48′N	11°17′
Nova Scotia, prov., Can. (skō′sh*à*)	60	44°28′N	65°00′
Novaya Sibir, i., Russia (nō′và-ya sė-bēr′)	5	75°00′N	149°00′
Novaya Zemlya, i., Russia (zĕm-lyà′)	5	72°00′N	54°46′
Novosibirsk, Russia (nô′vô-sė-bērsk′)	3	55°09′N	82°58′
Nubian Desert, des., Sudan (nōō′bĭ-*ă*n)	5	21°13′N	33°09′
Nunivak, i., Ak., U.S. (nōō′nĭ-văk)	60	60°25′N	167°42′
Nyasa, Lake, l., Afr. (nyä′sä)	5	10°45′S	34°30′

O

PLACE (Pronunciation)	PAGE	LAT.	LONG.
Oakland, Ca., U.S. (ōk′lănd)	60	37°48′N	122°16′
Oaxaca, Mex. (wä-hä′kä)	66	17°03′N	96°42′
Ob′, r., Russia	5	62°15′N	67°00′
Ocmulgee National Monument, rec., Ga., U.S. (ôk-mŭl′gē)	59	32°45′N	83°28′
Ohio, r., U.S. (ô′hī′ō)	61	37°25′N	88°05′
Okhotsk, Sea of, sea, Asia (ô-kôtsk′)	5	56°45′N	146°00′
Oklahoma City, Ok., U.S. (ô-klå-hō′m*à*)	43	35°27′N	97°32′
Olympic National Park, rec., Wa., U.S. (ô-lĭm′pĭk)	58	47°54′N	123°00′
Omaha, Ne., U.S. (ō′m*à*-hä)	60	41°18′N	95°57′
Omaha Indian Reservation, I.R., Ne., U.S.	58	42°09′N	96°08′
Oman, nation, Asia	3	20°00′N	57°45′
Oman, Gulf of, b., Asia	5	24°24′N	58°58′
Onezhskoye Ozero, Russia (ô-năsh′skô-yĕ ô′zĕ-rô)	5	62°02′N	34°35′
Ontario, Ca., U.S. (ŏn-tā′rĭ-ō)	43a	34°04′N	117°39′
Ontario, Lake, l., N.A.	61	43°35′N	79°05′
Orange, r., Afr. (ŏr′ĕnj)	5	29°15′S	17°30′
Organ Pipe Cactus National Monument, rec., Az., U.S. (ôr′g*ă*n pīp kăk′t*ŭ*s)	58	32°14′N	113°05′
Orinoco, r., Ven. (ô-rĭ-nō′kô)	4	8°32′N	63°13′
Orizaba, Pico de, vol., Mex. (ô-rē-zä′bä)	61	19°04′N	97°14′
Orlando, Fl., U.S. (ôr-lăn′dō)	43	28°32′N	81°22′
Ōsaka, Japan (ō′sä-kä)	3	34°40′N	135°27′
Oslo, Nor. (ŏs′lō)	3	59°56′N	10°41′
Ostrov, Russia (ôs-trôf′)	5	57°21′N	28°22′
Ottawa, Can. (ŏt′*à*-w*à*)	60	45°25′N	75°43′

P

PLACE (Pronunciation)	PAGE	LAT.	LONG.
Pacific Ocean, o.	4	0°00′	170°00′
Pacific Rim National Park, rec., Can.	58	49°00′N	126°00′
Padre Island, i., Tx., U.S. (pä′drā)	58	27°09′N	97°15′
Pakistan, nation, Asia	3	28°00′N	67°30′
Palau (Belau), nation, Oc. (pä-lä′ò)	3	7°15′N	134°30′
Palmas, Cape, c., Lib. (päl′mäs)	4	4°22′N	7°44′
Palmyra, i., Oc. (păl-mī′r*à*)	4	6°00′N	162°20′
Pamirs, mts., Asia	5	38°14′N	72°27′
Pampas, reg., Arg. (päm′päs)	4	37°00′S	64°30′
Panamá, Pan.	66	8°58′N	79°32′
Panamá, nation, N.A.	66	9°00′N	80°00′
Panamá, Istmo de, isth., Pan.	60	9°00′N	80°00′
Papua New Guinea, nation, Oc. (păp-ōō-*à*)(gĭne)	3	7°00′S	142°15′
Paraguay, nation, S.A. (păr′*à*-gwā)	2	24°00′S	57°00′
Paraguay, r., S.A. (pä-rä-gwä′y)	4	21°12′S	57°31′
Paraná, r., S.A.	4	24°00′S	54°00′
Pariñas, Punta, c., Peru (pōō′n-tä-pä-rē′n-yäs)	4	4°30′S	81°23′
Paris, Fr. (på-rē′)	3	48°51′N	2°20′
Parry, i., Can. (păr′ĭ)	60	45°15′N	80°00′
Pascua, Isla de (Easter Island), i., Chile	4	26°50′S	109°00′
Patagonia, reg., Arg. (păt-*à*-gō′nĭ-*à*)	4	46°45′S	69°30′
Peace River, Can. (pēs rĭv′ẽr)	61	56°14′N	117°17′

ăt; fin*ă*l; rāte; senåte; ärm; ȧsk; sof*à*; fâre; ch-choose; dh-as th in other; bē; ėvent; bĕt; recĕnt; cratẽr; g-gō; gh-guttural g; bĭt; ĭ-short neutral; rīde; ĸ-guttural k as ch in German ich;

PLACE (Pronunciation)	PAGE	LAT.	LONG.
Pelee, Point, c., Can. (pē'lē)	59	41°55'N	82°30'W
Penas, Golfo de, b., Chile (gôl-fô-dĕ-pĕ'n-äs)	4	47°15'S	77°30'W
Persian Gulf, b., Asia (pûr'zhăn)	5	27°38'N	50°30'E
Perth, Austl. (pûrth)	3	31°50'S	116°10'E
Peru, nation, S.A. (pĕ-rōō')	2	10°00'S	75°00'W
Peru-Chile Trench, deep	4	25°00'S	71°30'W
Philadelphia, Pa., U.S. (fĭl-ă-dĕl'phĭ-ă)	60	40°00'N	75°13'W
Philippines, nation, Asia (fĭl'ĭ-pēnz)	3	14°25'N	125°00'E
Philippine Trench, deep	5	10°30'N	127°15'E
Phoenix, Az., U.S. (fē'nĭks)	2	33°30'N	112°00'W
Phoenix Islands, is., Kir.	4	4°00'S	174°00'W
Pikes Peak, mtn., Co., U.S. (pīks)	61	38°49'N	105°03'W
Pipestone National Monument, rec., Mn., U.S. (pīp'stōn)	58	44°03'N	96°24'W
Pitcairn, dep., Oc.	2	25°04'S	130°05'W
Pittsburgh, Pa., U.S.	60	40°26'N	80°01'W
Plata, Río de la, est., S.A. (dālä plä'tä)	4	34°35'S	58°15'W
Platte, r., Ne., U.S. (plăt)	61	40°50'N	100°40'W
Poland, nation, Eur. (pō'lănd)	3	52°37'N	17°01'E
Popocatépetl Volcán, Mex. (pô-pô-kä-tä'pĕ't'l)	61	19°01'N	98°38'W
Port-au-Prince, Haiti (präɴs')	60	18°35'N	72°20'W
Portland, Me., U.S. (pōrt'lănd)	59	43°40'N	70°16'W
Portland, Or., U.S.	60	45°31'N	122°41'W
Porto Alegre, Braz. (pōr'tō ä-lā'grĕ)	2	29°58'S	51°11'W
Portugal, nation, Eur. (pōr'tu-găl)	2	38°15'N	8°08'W
Pretoria, S. Afr. (prĕ-tō'rĭ-ă)	3	25°43'S	28°16'E
Pribilof Islands, is., Ak., U.S. (prĭ'bĭ-lof)	60	57°00'N	169°20'W
Prince Edward Islands, is., S. Afr. (prĭns ĕd'wērd)	5	46°36'S	37°57'E
Prince Rupert, Can. (roo'pĕrt)	60	54°19'N	130°19'W
Providence, R.I., U.S. (prŏv'ĭ-dĕns)	43	41°50'N	71°23'W
Puebla, Mex. (pwā'blä)	6	19°02'N	98°11'W
Puerto Rico, dep., N.A. (pwĕr'tō rē'kō)	66	18°16'N	66°50'W
Puerto Rico Trench, deep	61	19°45'N	66°30'W
Pune, India	7	18°38'N	73°53'E
Purús, r., S.A. (pōō-rōō's)	4	6°45'S	64°34'W
Pusan, Kor., S.	7	35°08'N	129°05'E
P'yŏngyang, Kor., N. (pyŭng'găng')	7	39°03'N	125°48'E
Pyramid, l., Nv., U.S. (pĭ'ră-mĭd)	58	40°02'N	119°50'W
Pyramid Lake Indian Reservation, I.R., Nv., U.S.	58	40°17'N	119°52'W

Q

PLACE (Pronunciation)	PAGE	LAT.	LONG.
Qatar, nation, Asia (kä'tar)	3	25°00'N	52°45'E
Qausuittuq (Resolute), Can.	60	74°41'N	95°00'W
Québec, Can. (kwē-bĕk') (kā-bĕk')	60	46°49'N	71°13'W
Québec, prov., Can.	60	51°07'N	70°25'W
Queen Elizabeth Islands, is., Can. (ē-lĭz'ā-bĕth)	60	78°20'N	110°00'W
Quinault Indian Reservation, I.R., Wa., U.S.	58	47°27'N	124°34'W
Quito, Ec. (kē'tō)	2	0°17'S	78°32'W

R

PLACE (Pronunciation)	PAGE	LAT.	LONG.
Race, Cape, c., Can. (rās)	60	46°40'N	53°10'W
Rainier, Mount, mtn., Wa., U.S. (rā-nēr')	4	46°52'N	121°46'W
Raleigh, N.C., U.S.	43	35°45'N	78°39'W
Rangoon (Yangon), Mya. (răŋ-gōōn')	3	16°46'N	96°09'E
Rapid City, S.D., U.S.	42	44°06'N	103°14'W
Ras Dashen Terara, mtn., Eth. (räs dä-shän')	5	12°49'N	38°14'E
Recife, Braz. (rā-sē'fĕ)	2	8°09'S	34°59'W
Red, r., N.A. (rĕd)	61	48°00'N	97°00'W
Red, r., U.S.	61	31°40'N	92°55'W
Red Lake Indian Reservation, I.R., Mn., U.S.	59	48°09'N	95°55'W
Red Sea, sea	5	23°15'N	37°00'E
Redwood National Park, rec., Ca., U.S. (rĕd' wŏŏd)	58	41°20'N	124°00'W
Regina, Can. (rē-jī'nā)	60	50°25'N	104°39'W
Reindeer, l., Can. (rān'dēr)	61	57°36'N	101°23'W
Reno, Nv., U.S. (rē'nō)	43	39°32'N	119°49'W
Réunion, dep., Afr. (rā-ü-nyôn')	3	21°06'S	55°36'E
Revillagigedo, Islas, is., Mex. (ē's-läs-rĕ-vēl-yä-hĕ'gĕ-dō)	60	18°45'N	111°00'W
Reyes, Point, c., Ca., U.S.	58	38°00'N	123°00'W
Reykjavík, Ice. (rā'kyä-vēk)	60	64°09'N	21°39'W
Richmond, Va., U.S. (rĭch'mŭnd)	60	37°35'N	77°30'W
Riding Mountain National Park, rec., Can. (rīd'ĭng)	58	50°59'N	99°19'W
Rio de Janeiro, Braz. (rē'ō dā zhä-nā'ē-rō)	2	22°50'S	43°20'W
Riogrande, Tx., U.S. (rē'ō grän-dā)	61	26°23'N	98°48'W
Riverside, Ca., U.S. (rĭv'ĕr-sīd)	6	33°59'N	117°21'W
Riyadh, Sau. Ar.	3	24°31'N	46°47'E
Rochester, N.Y., U.S. (rŏch'ĕs-tēr)	43	43°15'N	77°35'W
Rocky Boys Indian Reservation, I.R., Mt., U.S.	58	48°08'N	109°34'W
Rocky Mountain National Park, rec., Co., U.S.	58	40°29'N	106°06'W
Rocky Mountains, mts., N.A.	61	50°00'N	114°00'W
Rogue, r., Or., U.S. (rōg)	58	42°32'N	124°13'W
Romania, nation, Eur. (rō-mā'nē-ā)	3	46°18'N	22°53'E
Rome (Roma), Italy	3	41°52'N	12°37'E
Rosario, Arg. (rō-zä'rē-ō)	2	32°58'S	60°42'W
Rosebud Indian Reservation, I.R., S.D., U.S. (rōz'bŭd)	58	43°13'N	100°42'W
Ross Sea, sea, Ant.	4	76°00'S	178°00'W
Rouyn, Can. (rōōn)	59	48°22'N	79°03'W
Russia, nation, Eur., Asia	3	61°00'N	60°00'E
Rwanda, nation, Afr.	68	2°10'S	29°37'E

S

PLACE (Pronunciation)	PAGE	LAT.	LONG.
Sable, Cape, c., Can. (sā'b'l)	60	43°25'N	65°24'W
Sable, Cape, c., Fl., U.S.	60	25°12'N	81°10'W
Sacramento, Ca., U.S. (săk-rā-mĕn'tō)	43	38°35'N	121°30'W
Sahara, des., Afr.	5	23°44'N	1°40'W
Saint Elias, Mount, mtn., N.A. (sānt ē-lī'ăs)	61	60°25'N	141°00'W
Saint Helena, i., St. Hel.	4	16°01'S	5°16'W
Saint John's, Can. (sānt jŏns)	60	47°34'N	52°43'W
Saint Kitts and Nevis, nation, N.A. (sānt kĭtts)(nē'vĭs)	66	17°24'N	63°30'W
Saint Lawrence, i., Ak., U.S. (sānt lô'rĕns)	60	63°10'N	172°12'W
Saint Lawrence, r., N.A.	61	48°24'N	69°30'W
Saint Lawrence, Gulf of, b., Can.	61	48°00'N	62°00'W
Saint Louis, Mo., U.S. (sānt lōō'ĭs) (lōō'ē)	60	38°39'N	90°15'W
Saint Lucia, nation, N.A.	66	13°54'N	60°40'W
Sainte Marie, Cap, c., Madag. (săɴ'tĕ-mä-rē')	3	25°31'S	45°00'E
Saint Paul, Mn., U.S. (sānt pôl')	60	44°57'N	93°05'W
Saint Paul, Île, i., Afr.	5	38°43'S	77°31'E
Saint Petersburg (Sankt-Peterburg) (Leningrad), Russia	3	59°57'N	30°20'E
Saint Petersburg, Fl., U.S. (pē'tērz-bûrg)	43	27°47'N	82°38'W
Saint Vincent and the Grenadines, nation, N.A. (vĭn'sĕnt)	66	13°20'N	60°50'W
Sakhalin, i., Russia (sā-kā-lēn')	5	52°00'N	143°00'E
Sala y Gómez, Isla, i., Chile	4	26°50'S	105°50'W
Salt Lake City, Ut., U.S. (sôlt lāk sĭ'tĭ)	60	40°45'N	111°52'W
Salvador (Bahia), Braz. (säl-vä-dōr') (bä-ē'ä)	2	12°59'S	38°27'W
Samoa, nation, Oc.	2	14°30'S	172°00'W
San Ambrosio, Isla, i., Chile (ē's-lä-dĕ-sän äm-brō'zē-ō)	4	26°40'S	80°00'W
San Antonio, Tx., U.S. (săn ăn-tō'nē-ō)	60	29°25'N	98°30'W
San Bernardino Mountains, mts., Ca., U.S. (bûr-när-dē'nō)	58	34°05'N	116°23'W
San Carlos Indian Reservation, I.R., Az., U.S. (săn kär'lōs)	58	33°27'N	110°15'W
San Diego, Ca., U.S. (săn dē-ā'gō)	6	32°43'N	117°10'W
San Félix, Isla, i., Chile (ē's-lä-dĕ-sän fā-lēks')	4	26°20'S	80°10'W
San Francisco, Ca., U.S. (săn frăn'sĭs'kō)	60	37°45'N	122°26'W
San Jorge, Golfo, b., Arg. (gôl-fō-sän-ĸō'r-ĸĕ)	4	46°15'S	66°45'W
San José, C.R. (sän hō-sā')	66	9°57'N	84°05'W
San Jose, Ca., U.S. (săn hō-zā')	43	37°20'N	121°54'W
San Juan, P.R. (sän hwän')	60	18°30'N	66°10'W
San Lucas, Cabo, c., Mex.	60	22°45'N	109°45'W
San Matías, Golfo, b., Arg. (sän mä-tē'äs)	4	41°30'S	63°45'W
San Salvador, El Sal. (sän säl-vä-dōr')	66	13°45'N	89°11'W
San Salvador (Watling), i., Bah. (săn săl'vă-dôr)	60	24°05'N	74°30'W
Santa Ana, Ca., U.S. (săn'tă ăn'ă)	43	33°45'N	117°52'W
Santiago, Chile (sän-tē-ä'gō)	2	33°26'S	70°40'W
Santo Domingo, Dom. Rep. (săn'tō dō-mĭn'gō)	60	18°30'N	69°55'W
São Francisco, Braz. (souɴ fräɴ-sēsh'kō)	4	15°59'S	44°42'W
São Paulo, Braz. (souɴ' pou'lō)	2	23°34'S	46°38'W
São Roque, Cabo de, c., Braz. (kä'bo-dĕ-souɴ' rō'kĕ)	4	5°06'S	35°11'W
Sao Tome and Principe, nation, Afr. (prēn'sē-pĕ)	68	1°00'N	6°00'E
Sapporo, Japan (säp-pō'rō)	7	43°02'N	141°29'E
Sarasota, Fl., U.S. (săr-ă-sōtă)	53a	27°27'N	82°30'W
Sardinia, i., Italy (sär-dĭn'ĭă)	5	40°08'N	9°05'E
Saskatchewan, r., Can. (săs-kăch'ē-wän)	61	53°45'N	103°20'W
Saskatoon, Can. (săs-kă-tōōn')	58	52°07'N	106°38'W
Saudi Arabia, nation, Asia (sā-ō'dĭ ā-rā'bĭ-ă)	3	22°40'N	46°00'E
Savannah, Ga., U.S. (să-văn'ă)	60	32°04'N	81°07'W
Scandinavian Peninsula, pen., Eur.	5	62°00'N	14°00'E
Scranton, Pa., U.S. (skrăn'tŭn)	53	41°15'N	75°45'W
Seattle, Wa., U.S. (sē-ăt''l)	60	47°36'N	122°20'W
Sénégal, nation, Afr. (sĕn-ĕ-gôl')	68	14°53'N	14°58'W
Seoul (Sŏul), Kor., S.	3	37°35'N	127°03'E
Sequoia National Park, rec., Ca., U.S. (sē-kwoi'ă)	58	36°34'N	118°37'W
Serbia and Montenegro (Yugoslavia), nation, Eur.	3	44°00'N	21°00'E
Seward, Ak., U.S. (sū'ărd)	60	60°18'N	149°28'W
Seychelles, nation, Afr. (sā-shĕl')	68	5°20'S	55°10'E
Shanghai, China (shăng'hī')	3	31°14'N	121°27'E
Shark Bay, b., Austl. (shärk)	5	25°30'S	113°00'E
Shasta, Mount, mtn., Ca., U.S. (shăs'tă)	61	41°35'N	122°12'W
Shenyang, China (shŭn-yäŋ)	3	41°45'N	123°22'E
Shetland Islands, is., Scot., U.K. (shĕt'lănd)	5	60°35'N	2°10'W
Sicily, i., Italy (sĭs'ĭ-lē)	5	37°38'N	13°30'E
Sierra Leone, nation, Afr. (sē-ĕr'rä lā-ō'nā)	68	8°48'N	12°30'W
Sikhote Alin', Khrebet, mts., Russia (se-kô'ta a-lēn')	5	45°00'N	135°45'E
Singapore, nation, Asia (sĭn'gā-pōr')	3	1°22'N	103°45'E
Sioux Falls, S.D., U.S. (sōō fôlz)	58	43°33'N	96°43'W
Sitka, Ak., U.S. (sĭt'kā)	60	57°08'N	135°18'W
Slovakia, nation, Eur.	3	48°50'N	20°00'E
Slovenia, nation, Eur.	3	45°58'N	14°43'E
Snake, r., U.S. (snāk)	61	45°30'N	117°00'W
Society Islands, is., Fr. Poly. (sō-sī'ĕ-tē)	4	15°00'S	157°30'W
Solomon Islands, nation, Oc. (sŏl'ō-mŭn)	3	7°00'S	160°00'E
Somalia, nation, Afr. (sō-mä'lē-ă)	68a	3°28'N	44°47'E
Sources, Mount aux, mtn., Afr. (môɴ'tō sōrs')	5	28°47'S	29°04'E
South Africa, nation, Afr.	68	28°00'S	24°50'E
South America, cont.	62	15°00'S	60°00'W
South China Sea, sea, Asia (chī'nā)	5	15°23'N	114°12'E
South East Cape, c., Austl.	5	43°47'S	146°03'E
South Georgia, i., S. Geor. (jôr'jā)	4	54°00'S	37°00'W
South Island, i., N.Z.	5	42°40'S	169°00'E
South Orkney Islands, is., Ant.	4	57°00'S	45°00'W
South Pole, pt. of. i., Ant.	4	90°00'S	0°00'
South Sandwich Islands, is., S. Geor. (sănd'wĭch)	4	58°00'S	27°00'W
South Shetland Islands, is., Ant.	4	62°00'S	70°00'W
Spain, nation, Eur. (spān)	3	40°15'N	4°30'W
Spencer Gulf, b., Austl. (spĕn'sēr)	5	34°20'S	136°55'E
Spokane, Wa., U.S. (spōkăn')	60	47°39'N	117°25'W
Spokane Indian Reservation, I.R., Wa., U.S.	58	47°55'N	118°00'W
Springfield, Ma., U.S. (sprĭng'fēld)	53	42°05'N	72°35'W
Sri Lanka, nation, Asia	3	8°45'N	82°30'E
Standing Rock Indian Reservation, I.R., N.D., U.S. (stănd'ĭng rŏk)	58	47°07'N	101°05'W
Stanovoy Khrebet, mts., Russia (stŭn-ā-voi')	5	56°12'N	127°12'E
Stewart, Island, i., N.Z. (stū'ērt)	5	46°56'N	167°48'W
Stockbridge Munsee Indian Reservation, I.R., Wi., U.S. (stŏk'brĭdj mŭn-sē)	59	44°49'N	89°00'W
Stockholm, Swe. (stŏk'hōlm)	3	59°23'N	18°00'E
Stockton, Ca., U.S. (stŏk'tŭn)	53	37°56'N	121°16'W
Stuttgart, Ger. (shtōōt'gärt)	6	48°48'N	9°15'E
Sucre, Bol. (sōō'krā)	2	19°06'S	65°16'W
Sudan, nation, Afr. (sōō-dän')	68	14°00'N	28°00'E
Sudbury, Can. (sŭd'bēr-ē)	59	46°28'N	81°00'W
Sulu Sea, sea, Asia	5	8°25'N	119°00'E
Sumatera, (Sumatra) i., Indon. (sō-mä-trā)	3	2°06'N	99°40'E
Sumatra *see* Sumatera, i., Indon.	3	2°06'N	99°40'E
Sunda, Selat, strt., Indon.	5	5°45'S	106°15'E
Superior, Lake, l., N.A. (su-pē'rĭ-ēr)	61	47°38'N	89°20'W
Surabaya, Indon.	7	7°23'S	112°45'E
Surat, India (sō'rŭt)	7	21°08'N	73°22'E
Suriname, nation, S.A. (sōō-rē-näm')	2	4°00'N	56°00'W
Suva, Fiji	3	18°08'S	178°25'E
Svalbard (Spitsbergen), dep., Nor. (sväl'bärt) (spĭts'bûr-gĕn)	3	77°00'N	20°00'E
Swaziland, nation, Afr. (Swä'zē-lănd)	68	26°45'S	31°30'E
Sweden, nation, Eur. (swē'dĕn)	3	60°10'N	14°10'E
Switzerland, nation, Eur. (swĭt'zēr-lănd)	3	46°30'N	7°43'E
Sydney, Austl. (sĭd'nē)	3	33°55'S	151°17'E
Syracuse, N.Y., U.S. (sĭr'ă-kūs)	53	43°05'N	76°10'W
Syr Darya, r., Asia	5	44°15'N	65°45'E
Syria, nation, Asia (sĭr'ĭ-ā)	3	35°00'N	37°15'E
Syrian Desert, des., Asia	5	32°00'N	40°00'E

T

PLACE (Pronunciation)	PAGE	LAT.	LONG.
Tabuaeran, i., Kir.	4	3°52'N	159°20'W
Tacoma, Wa., U.S. (tă-kō'mă)	43	47°14'N	122°27'W
Taegu, Kor., S. (tī'gōō')	7	35°49'N	128°41'E
Tahiti, i., Fr. Poly. (tä-hē'tē) (tä'ē-tē')	4	17°30'S	149°30'W
Tahoe, l., U.S. (tä'hō)	58	39°09'N	120°18'W
T'aipei, Tai. (tī'pā')	7	25°02'N	121°38'E
Taiwan (Formosa), nation, Asia (tī-wän) (fŏr-mō'să)	3	23°30'N	122°20'E
Tajikistan, nation, Asia	3	39°22'N	69°30'E
Tallahassee, Fl., U.S. (tăl-ă-hăs'ē)	59	30°25'N	84°17'W
Tampa, Fl., U.S. (tăm'pā)	6	27°57'N	82°25'W
Tampico, Mex. (täm-pē'kō)	60	22°14'N	97°51'W
Tanana, Ak., U.S. (tă'nă-nō)	60	65°18'N	152°20'W
Tanganyika, Lake, l., Afr.	5	5°15'S	29°40'E
Tanzania, nation, Afr.	68	6°48'S	33°58'E
Tarim Basin, basin, China (tă-rĭm')	5	39°52'N	82°34'E
Tashkent, Uzb. (tăsh'kĕnt)	3	41°23'N	69°04'E
Tasmania, state, Austl.	5	41°28'S	142°30'E
Tasman Sea, sea, Oc.	5	29°30'S	155°00'E
Taymyr, Poluostrov, pen., Russia (tī-mĭr')	5	75°15'N	95°00'E
Tegucigalpa, Hond. (tă-gōō-sē-gäl'pä)	66	14°08'N	87°15'W
Tehrān, Iran (tĕ-hrän')	3	35°45'N	51°30'E
Tehuantepec, Istmo de, isth., Mex. (ĕ'st-mō dĕ)	61	17°55'N	94°35'W

ng-sing; ŋ-baŋk; ɴ-nasalized n; nŏd; cŏmmit; ōld; ôbey; ôrder; oi-boil; fōōd; ŏŏ-as oo in foot; ou-out; s-soft; sh-dish; th-thin; pūre; ůnite; ûrn; stŭd; circŭs; ü-as in French tu; '-indeterminate vowel.

PLACE (Pronunciation)	PAGE	LAT.	LONG.
Thailand, nation, Asia	3	16°30′N	101°00′E
Thailand, Gulf of, b., Asia	5	11°37′N	100°46′E
Theodore Roosevelt National Park, rec., N.D., U.S. (thē-ŏ-doŕ rōō-sà-vĕlt)	58	47°20′N	103°42′W
Thule, Grnld.	60	76°34′N	68°47′W
Thunder Bay, Can. (thŭn′dĕr)	59	48°28′N	89°12′W
Tianjin, China (tĕn-jyĭn)	7	39°08′N	117°14′E
Tien Shan, mts., Asia	5	42°00′N	78°46′E
Tierra del Fuego, i., S.A. (tyĕr′rä dĕl fwä′gŏ)	4	53°50′S	68°45′W
Timor, i., Asia (tē-mōr′)	5	10°08′S	125°00′E
Timor Sea, sea	5	12°40′S	125°00′E
Titicaca, Lago, l., S.A. (lä′gô-tē-tē-kä′kä)	4	16°12′S	70°33′W
Tobago, i., Trin. (tô-bä′gŏ)	66	11°15′N	60°30′W
Tocantins, r., Braz. (tō-kän-tēns′)	4	3°28′S	49°22′W
Togo, nation, Afr. (tō′gō)	68	8°00′N	0°52′E
Tohono O'odham Indian Reservation, I.R., Az., U.S.	58	32°33′N	112°12′W
Tokelau, dep., Oc. (tō-kĕ-lä′ô)	2	8°00′S	176°00′W
Tōkyō, Japan (tō′kyō)	3	35°42′N	139°46′E
Toledo, Oh., U.S.	53	41°40′N	83°35′W
Tonga, nation, Oc. (tŏŋ′gà)	2	18°50′S	175°20′W
Tonga Trench, deep	2	23°00′S	172°30′W
Tonto National Monument, rec., Az., U.S. (tŏn′tŏ)	58	33°33′N	111°08′W
Toronto, Can. (tô-rŏn′tō)	60	43°40′N	79°23′W
Torres Strait, strt., Austl. (tôr′rĕs)	5	10°30′S	141°30′E
Toubkal, Jebel, mtn., Mor.	4	31°15′N	7°46′W
Trinidad and Tobago, nation, N.A. (trĭn′ĭ-dăd) (tô-bä′gŏ)	66	11°00′N	61°00′W
Tripoli (Tarabulus), Libya	3	32°50′N	13°13′E
Tristan da Cunha Islands, is., St. Hel. (trĕs-tän′dä kōōn′yà)	4	35°30′S	12°15′W
Tuamoto, Îles, Fr. Poly. (tōō-ä-mô′tōō)	4	19°00′S	141°20′W
Tucson, Az., U.S. (tōō-sŏn′)	43	32°15′N	111°00′W
Tuktut Nogait National Park, rec., Can.	58	69°00′N	122°00′W
Tule River Indian Reservation, I.R., Ca., U.S. (tōō′lå)	58	36°00′N	118°40′W
Tulsa, Ok., U.S. (tŭl′sà)	43	36°08′N	95°58′W
Tunisia, nation, Afr. (tu-nĭzh′ē-à)	68	35°00′N	10°11′E
Turkey, nation, Asia	3	38°45′N	32°00′E
Turkmenistan, nation, Asia	3	40°46′N	56°01′E
Turks, is., T./C. Is. (tûrks)	66	21°40′N	71°45′W
Turtle Mountain Indian Reservation, I.R., N.D., U.S.	58	48°45′N	99°57′W
Tutuila, i., Am. Sam.	4	14°18′S	170°42′W

U

Ubangi, r., Afr. (ōō-bäŋ′gē)	5	3°00′N	18°00′E
Uganda, nation, Afr. (ōō-gän′dä) (û-gän′dà)	68	2°00′N	32°28′E
Uintah and Ouray Indian Reservation, I.R., Ut., U.S.	58	40°20′N	110°20′W
Ukraine, nation, Eur.	3	49°15′N	30°15′E
Ulan Bator (Ulaanbaatar), Mong.	3	47°56′N	107°00′E
Umatilla Indian Reservation, I.R., Or., U.S. (ū-mà-tĭl′à)	58	45°38′N	118°35′W
Uncompahgre Plateau, plat., Co., U.S. (ŭn-kŭm-pä′grĕ)	58	38°40′N	108°40′W
Ungava, Péninsule d', pen., Can.	60	59°55′N	74°00′W
Ungava Bay, b., Can. (ŭŋ-gá′và)	61	59°46′N	67°18′W
United Kingdom, nation, Eur.	3	56°30′N	1°40′W

PLACE (Pronunciation)	PAGE	LAT.	LONG.
United States, nation, N.A.	60	38°00′N	110°00′W
Urals, mts., Russia	5	56°28′N	58°13′E
Uruguay, nation, S.A. (ōō-rōō-gwī′) (ū′rōō-gwā)	2	32°45′S	56°00′W
Ürümqi, China (û-rŭm-chyē)	3	43°49′N	87°43′E
Ute Mountain Indian Reservation, I.R., N.M., U.S.	58	36°57′N	108°34′W
Uzbekistan, nation, Asia	3	42°42′N	60°00′E

V

Valparaíso, Chile (väl′pä-rä-ē′sŏ)	2	33°02′S	71°32′W
Vancouver, Can. (văn-kōō′vĕr)	60	49°16′N	123°06′W
Vancouver Island, i., Can.	60	49°50′N	125°05′W
Vanuatu, nation, Oc.	3	16°02′S	169°15′E
Venezuela, nation, S.A. (vĕn-ê-zwē′là)	2	8°00′N	65°00′W
Venezuela, Golfo de, b., S.A. (gôl-fō-dĕ)	61	11°34′N	71°02′W
Ventura, Ca., U.S. (vĕn-tōō′rà)	53	34°18′N	119°18′W
Veracruz, Mex. (vā-rä-krōōz′)	60	19°13′N	96°07′W
Vert, Cap, c., Sen.	4	14°43′N	17°30′W
Victoria, Can. (vĭk-tō′rĭ-à)	58	48°26′N	123°23′W
Victoria, l., Afr.	5	0°50′S	32°50′E
Victoria Island, i., Can.	60	70°13′N	107°45′W
Victoria Land, reg., Ant.	5	75°00′S	160°00′E
Vietnam, nation, Asia (vyĕt′näm′)	3	18°00′N	107°00′E
Virgin Islands, is., N.A. (vûr′jĭn)	66	18°15′N	64°00′W
Viscount Melville Sound, strt., Can.	61	74°00′N	110°00′W
Viti Levu, i., Fiji	5	18°00′S	178°00′E
Vladivostok, Russia (vlà-dē-vôs-tŏk′)	3	43°06′N	131°47′E
Volga, r., Russia (vôl′gä)	5	47°30′N	46°20′E
Voyageurs National Park, rec., Mn., U.S.	59	48°30′N	92°40′W
Vrangelya (Wrangel), i., Russia	5	71°25′N	178°30′W
Vuntut National Park, rec., Can.	58	68°27′N	139°58′W

W

Wake, i., Oc. (wăk)	3	19°25′N	167°00′E
Walker River Indian Reservation, I.R., Nv., U.S.	58	39°06′N	118°20′W
Warm Springs Indian Reservation, I.R., Or., U.S. (wôrm sprĭnz)	58	44°55′N	121°30′W
Warsaw, Pol. (wôr′sô)	6	52°15′N	21°05′E
Washington, D.C., U.S. (wôsh′ĭng-tŭn)	60	38°50′N	77°00′W
Waterton-Glacier International Peace Park, rec., N.A. (wô′ter-tŭn-glā′shŭr)	58	48°55′N	114°10′W
Waterton Lakes National Park, rec., Can.	58	49°05′N	113°50′W
Weddell Sea, sea, Ant. (wĕd′ĕl)	4	73°00′S	45°00′W
Wellington, N.Z. (wĕl′lĭng-tŭn)	3	41°15′S	174°45′E
Western Sahara, dep., Afr. (sá-hä′rà)	68	23°05′N	15°33′W
West Indies, is., (ĭn′dēz)	60	19°00′N	78°30′W
West Palm Beach, Fl., U.S. (päm bēch)	43	26°44′N	80°04′W
White Earth Indian Reservation, I.R., Mn., U.S.	58	47°18′N	95°42′W
Whitehorse, Can. (whīt′hôrs)	60	60°39′N	135°01′W
White Mountains, mts., Me., U.S.	59	44°22′N	71°15′W
White Mountains, mts., N.H., U.S.	59	42°20′N	71°05′W
White Sands National Monument, rec., N.M., U.S.	58	32°50′N	106°20′W
White Sea, sea, Russia	5	66°00′N	40°00′E

PLACE (Pronunciation)	PAGE	LAT.	LONG.
Whitney, Mount, mtn., Ca., U.S.	61	36°34′N	118°18
Wichita, Ks., U.S. (wĭch′ĭ-tô)	60	37°42′N	97°21
Wilkes Land, reg., Ant.	5	71°00′S	126°00
Wind River Indian Reservation, I.R., Wy., U.S.	58	43°26′N	109°00
Windward Islands, is., N.A. (wĭnd′wĕrd)	4	12°45′N	61°40
Windward Passage, strt., N.A.	61	19°30′N	74°20
Winnebago Indian Reservation, I.R., Ne., U.S.	58	42°15′N	96°06
Winnipeg, Can. (wĭn′ĭ-pĕg)	60	49°53′N	97°09
Winnipeg, Lake, l., Can.	61	52°00′N	97°00
Woods, Lake of the, l., N.A. (wŏdz)	61	49°25′N	93°25
Worcester, Ma., U.S. (wŏs′tĕr)	53	42°16′N	71°49
Wrangell-Saint Elias National Park, rec., Ak., U.S.	58	61°00′N	142°00
Wuhan, China	3	30°30′N	114°15

X

Xi'an, China (shyē-än)	7	34°20′N	109°00

Y

Yakima Indian Reservation, I.R., Wa., U.S. (yăk′ĭ-mà)	58	46°16′N	121°03
Yangtze (Chang), r., China (yäng′tse) (chäŋ)	5	30°30′N	117°25
Yap, i., Micron. (yăp)	5	11°00′N	138°00
Yellow Sea, sea, Asia (yĕl′ŏ)	5	35°20′N	122°15
Yellowstone, r., U.S.	61	46°00′N	108°00
Yellowstone National Park, rec., U.S. (yĕl′ŏ-stōn)	58	44°45′N	110°35
Yemen, nation, Asia (yĕm′ĕn)	3	15°00′N	47°00
Yenisey, r., Russia (yĕ-nĕ-sĕ′ĕ)	5	71°00′N	82°00
Yokohama, Japan (yō′kô-hä′mà)	7	35°37′N	139°40
York, Cape, c., Austl. (yôrk)	5	10°45′S	142°35
York, Kap, c., Grnld.	60	75°30′N	73°00
Yosemite National Park, rec., Ca., U.S. (yŏ-sĕm′ĭ-tĕ)	58	38°03′N	119°36
Youngstown, Oh., U.S.	53	41°05′N	80°40
Yucatán Channel, strt., N.A. (yōō-kä-tän′)	61	22°30′N	87°00
Yucatán Peninsula, pen., N.A.	60	19°30′N	89°00
Yukon, r., N.A. (yōō′kŏn)	61	64°00′N	159°30

Z

Zambezi, r., Afr. (zăm-bä′zĕ)	5	16°00′S	29°45
Zambia, nation, Afr. (zăm′bē-à)	68	14°23′S	24°15
Zanzibar, i., Tan. (zăn′zĭ-bär)	5	6°20′S	39°37
Zemlya Frantsa-Iosifa (Franz Josef Land), is., Russia	5	81°32′N	40°00
Zimbabwe, nation, Afr. (rô-dē′zhĭ-à)	68	17°50′S	29°30
Zion National Park, rec., Ut., U.S.	58	37°20′N	113°00
Zuni Indian Reservation, I.R., N.M., U.S. (zōō′nē)	58	35°10′N	108°40

SUBJECT INDEX

Listed below are major topics covered by the thematic maps, graphs and/or statistics.
Page citations are for world, continent and country maps and for world tables.

Abaca 22
Agriculture 14–15, 41, 50–51, 63
Air travel 43
Alcohol consumption 22
Alliances 38
Aluminum 28
Anthracite, see Coal
Apples 20
Aquifers 40, 52
Armed forces 37
Arms exports 37
Atmospheric pressure 47
Bananas 20
Barley 18
Bauxite 28, 41, 63
Beef 23
Beer 22
Birth rate 7, 8
Bituminous coal, see Coal
Calorie consumption 12
Canola oil 21
Cassava 19
Cattle 23
Chickens 24
Chromite 30
Cigarette consumption 20
Citrus fruit 20
Coal 34–35, 40, 41, 62, 63
Cobalt 30
Cocoa beans 18
Coconuts 21
Coffee 17
Commuting time 55
Conflicts 39
Contraception use 8
Copper 28, 41, 63
Corn 17, 21
Corn oil 21
Cotton 22
Cottonseed 21
Dates 20
Death rate 7, 8
Deciduous fruit 20
Deforestation 27
Desertification 40, 62, 68
Ducks 24
Earnings, see Income
Earth properties iii
Earthquakes 40, 62, 68
Economic alliances, see Alliances
Education 10, 55
Eggs 24
Electricity, see Energy
Energy . . . . 32, 33, 40, 62.
    Also see specific energy generation
        methods (geothermal, hydroelectric,
            nuclear, and thermal).
Environments 42
Ethnic groups, see Race and ethnicity
Exports 36
Federal lands 58–59
Fertilizers 25
Fisheries 20
Flax 22
Floods 40, 62
Food aid 25
Forest products 26–27
Forested lands 26–27
Frost 47
Fruit 20
Fuels 34–35
    Also see specific fuel types (coal,
        natural gas, petroleum, and uranium)

GDP, see Gross Domestic Product
Geothermal energy 32, 40
Glaciation 40, 44–45, 46
Gold 31
Grain sorghum 18
Grapes 22
Gross Domestic Product 10
Ground water 52
Hazardous waste sites 52
Highways 58–59
HIV infection 13
Hops 22
Housing 55
Hydroelectric energy 32, 40, 62
Imports 36
Income 56
Indian reservations 58–59
Infant mortality 8
Interstate highways 58–59
Iron ore 30, 41, 63
Jute 22
Kapok 22
Labor force 57
Land areas 2–3
Land elevations 4–5
Landforms 40, 44–45, 62, 68
Languages 11, 56, 64–65
Lead 29, 41, 63
Life expectancy 9
Lignite, see Coal
Literacy 10
Maize, see Corn
Manganese 30
Manufacturing 57
Map legend 1
Map projections . . . . vi–viii
Map scale v
Map symbols, see Map legend
Military alliances, see Alliances
Military expenditures 37
Military installations 58–59
Millet 18
Minerals 41, 63
    Also see specific minerals.
Moisture regions 47
Molybdenum 30
National Parks 58–59
Natural gas 34–35, 40, 62
Natural hazards 40, 62, 68
Natural increase 9
Natural vegetation, see Vegetation
Nickel 30, 41
Nuclear energy 32, 40, 62
Nuclear weapons 37
Nutrition 12, 25
Oats 17
Ocean depths 4–5
Oil, see Petroleum
Oil palm fruit 21
Olive oil 21
Olives 21
Palm oil 21
Peacekeeping operations 39
Peanut oil 21
Peanuts 21
Petroleum 34–35, 40, 41, 62, 63
Physicians 13
Physiography 44–45
Pigs 23
Pineapples 20
Platinum 31
Political alliances, see Alliances
Political change, Africa 68

Population 2–3, 6–7, 53, 54, 55
Population change 55
Population density 6–7, 41, 53, 63
Population growth, see Natural increase
Population, age 55
Population, age-sex composition 7
Population, ethnic, see Race and ethnicity
Population, urban, see Urban population
Population, world 3, 9
Pork 23
Potatoes 19
Poultry 24
Poverty 56
Precipitation 41, 46, 63
Protein consumption 12
Race and ethnicity 54, 62, 66, 67, 68
Railroads 43
Rapeseed 21
Refugees 39
Religions 11
Rice 18
Rubber 22
Rye 16
Sea ice 40, 62
Sheep 24
Silver 31
Sisal 22
Soybean oil 21
Soybeans 21
Steel 31
Sugar 19
Sunflower oil 21
Sunflower seeds 21
Sunshine 47
Tea 16
Territorial evolution 56
Thematic map types ix
Thermal efficiency 47
Thermal energy 32
Time zones 43
Tin 28, 63
Tobacco 20
Tornadoes 40
Transportation 43, 58–59
Tropical storms 40, 68
Tsunamis 40, 62, 68
Tungsten 30, 63
Turkeys 24
Unemployment 56
Uranium 34–35, 40, 62
Urban population 12, 53
Vanadium 30
Veal 23
Vegetable oils 21
Vegetation 41, 48–49, 63
Volcanoes 40, 62, 68
Water quality 52
Water resources 40, 52
Waterways 43
Wheat 16
Wine 22
Women heads of household 56
Women's economic activity rate 37
Women's legislative participation rate 37
Women's voting rights 37
Wood, see Forest products
Wool 24
Zinc 29, 41, 63

SOURCES

The following sources have been consulted during the process of creating and updating the thematic maps and statistics for the 21st Edition.

Air Carrier Traffic at Canadian Airports, Statistics Canada
Annual Coal Report, U.S. Dept. of Energy, Energy Information Administration
Armed Conflicts Report, Project Ploughshares
Atlas of Canada, Natural Resources Canada
Canadian Minerals Yearbook, Statistics Canada
Census of Canada, Statistics Canada
Census of Population, U.S. Census Bureau
Chromium Industry Directory, International Chromium Development Association
Coal Fields of the Conterminous United States, U.S. Geological Survey
Coal Quality and Resources of the Former Soviet Union, U.S. Geological Survey
Coal-Bearing Regions and Structural Sedimentary Basins of China and Adjacent Seas, U.S. Geological Survey
Commercial Service Airports in the United States with Percent Boardings Change, Federal Aviation Administration (FAA)
Completed Peacekeeping Operations, Center for Defense Information
Conventional Arms Transfers to Developing Nations, Library of Congress, Congressional Research Service
Current Status of the World's Major Episodes of Political Violence: Hot Wars and Hot Spots, Center for Systemic Peace
Dependencies and Areas of Special Sovereignty, U.S. Dept. of State, Bureau of Intelligence and Research
Earth's Seasons—Equinoxes, Solstices, Perihelion, and Aphelion, U.S. Naval Observatory
EarthTrends: The Environmental Information Portal, World Resources Institute and World Conservation Monitoring Centre 2003. Available at http://earthtrends.wri.org/ Washington, D.C.: World Resources Institute
Economic Census, U.S. Census Bureau
Employment, Hours, and Earnings from the Current Employment Statistics Survey, U.S. Dept. of Labor, Bureau of Labor Statistics
Energy Statistics Yearbook, United Nations Dept. of Economic and Social Affairs
Epidemiological Fact Sheets by Country, Joint United Nations Program on HIV/AIDS (UNAIDS), World Health Organization, United Nations Children's Fund (UNICEF)
Estimated Water Use in the United States, U.S. Geological Survey
Estimates of Health Personnel, World Health Organization
FAO Food Balance Sheet, Food and Agriculture Organization of the United Nations (FAO)
FAO Statistical Databases (FAOSTAT), Food and Agriculture Organization of the United Nations (FAO)
Fishstat Plus, Food and Agriculture Organization of the United Nations (FAO)
Geothermal Resources Council Bulletin, Geothermal Resources Bulletin
Geothermal Resources in China, Bob Lawrence and Associates, Inc.
Global Alcohol Database, World Health Organization
Global Forest Resources Assessment, Food and Agriculture Organization of the United Nations (FAO), Forest Resources Assessment Programme
Great Lakes Factsheet Number 1, U.S. Environmental Protection Agency
The Hop Atlas, Joh. Barth & Sohn GmbH & Co. KG
Human Development Report 2003, United Nations Development Programme, © 2003 by United Nations Development Programme. Used by permission of Oxford University Press, Inc.
Installed Generating Capacity, International Geothermal Association
International Database, U.S. Census Bureau
International Energy Annual, U.S. Dept. of Energy, Energy Information Administration
International Journal on Hydropower and Dams, International Commission on Large Dams
International Petroleum Encyclopedia, PennWell Publishing Co.
International Sugar and Sweetener Report, F.O. Licht, Licht Interactive Data
International Trade Statistics, World Trade Organization
International Water Power and Dam Construction Yearbook, Wilmington Publishing
Iron and Steel Statistics, U.S. Geological Survey, Thomas D. Kelly and Michael D. Fenton
Lakes at a Glance, LakeNet
Land Scan Global Population Database, U.S. Dept. of Energy, Oak Ridge National Laboratory (© 2003 UT-Battelle, LLC. All rights reserved. Notice: These data were produced by UT-Battelle, LLC under Contract No. DE-AC05-00OR22725 with the Department of Energy. The Government has certain rights in this data. Neither UT-Battelle, LLC nor the United States Department of Energy, nor any of their employees, makes any warranty, express or implied, or assumes any legal liability or responsibility for the accuracy, completeness, or usefulness of any data, apparatus, product, or process disclosed, or represents that its use would not infringe privately owned rights.)
Largest Rivers in the United States, U.S. Geological Survey

Lengths of the Major Rivers, U.S. Geological Survey
Likely Nuclear Arsenals Under the Strategic Offensive Reductions Treaty, Center for Defense Information
Major Episodes of Political Violence, Center for Systemic Peace
Maps of Nuclear Power Reactors, International Nuclear Safety Center
Mineral Commodity Summaries, U.S. Geological Survey, Bureau of Mines
Mineral Industry Surveys, U.S. Geological Survey, Bureau of Mines
Minerals Yearbook, U.S. Geological Survey, Bureau of Mines
National Priorities List, U.S. Environmental Protection Agency
National Tobacco Information Online System (NATIONS), U.S. Dept. of Health and Human Services, Centers for Disease Control and Prevention (CDC)
Natural Gas Annual, U.S. Dept. of Energy, Energy Information Administration
New and Recent Conflicts of the World, The History Guy
Nuclear Power Reactors in the World, International Atomic Energy Agency
Oil and Gas Journal DataBook, PennWell Publishing Co.
Oil and Gas Resources of the World, Oilfield Publications, Ltd.
Petroleum Supply Annual, U.S. Dept. of Energy, Energy Information Administration
Population of Capital Cities and Cities of 100,000 and More Inhabitants, United Nations Dept. of Economic and Social Affairs
Preliminary Estimate of the Mineral Production of Canada, Natural Resources Canada
Red List of Threatened Species, International Union for Conservation and Natural Resources
Significant Earthquakes of the World, U.S. Geological Survey
State of Food Insecurity in the World, Food and Agriculture Organization of the United Nations (FAO)
State of the World's Children, United Nations Children's Fund (UNICEF)
Statistical Abstract of the United States, U.S. Census Bureau
Statistics on Asylum-Seekers, Refugees and Others of Concern to UNHCR, United Nations High Commissioner for Refugees (UNHCR)
Survey of Energy Resources, World Energy Council
Tables of Nuclear Weapons Stockpiles, Natural Resources Defense Council
TeleGeography Research, PriMetrica, Inc. (www.primetrica.com)
Tobacco Atlas, World Health Organization
Tobacco Control Country Profiles, World Health Organization
Transportation in Canada, Minister of Public Works and Government Services, Transport Canada
UNESCO Statistical Tables, United Nations Educational, Scientific and Cultural Organization (UNESCO)
United Nations Commodity Trade Statistics (COMTRADE), United Nations Dept. of Economic and Social Affairs
United Nations Peacekeeping in the Service of Peace, United Nations Dept. of Peacekeeping Operations
United Nations Peacekeeping Operations, United Nations Dept. of Peacekeeping Operations
Uranium: Resources, Production and Demand, United Nations Organization for Economic Co-operation and Development (OECD)
Volcanoes of the World, Smithsonian National Museum of Natural History
Water Account for Australia, Australian Bureau of Statistics
Women in National Parliaments, Inter-Parliamentary Union
Women's Suffrage, Inter-Parliamentary Union
The World at War, Center for Defense Information, The Defense Monitor
The World at War, Federation of American Scientists, Military Analysis Network
World Conflict List, National Defense Council Foundation
World Contraceptive Use, United Nations Dept. of Economic and Social Affairs
The World Factbook, U.S. Dept. of State, Central Intelligence Agency (CIA)
World Facts and Maps, Rand McNally
World Lakes Database, International Lake Environment Committee
World Population Prospects, United Nations Dept. of Economic and Social Affairs
World Urbanization Prospects, United Nations Dept. of Economic and Social Affairs
World Water Resources and Their Use, State Hydrological Institute of Russia/UNESCO
The World's Nuclear Arsenal, Center for Defense Information

Special Acknowledgements
The American Geographical Society, for permission to use the Miller cylindrical projection.
The Association of American Geographers, for permission to use R. Murphy's landforms map.
The McGraw-Hill Book Company, for permission to use G. Trewartha's climatic regions map.
The University of Chicago Press, for permission to use Goode's Homolosine equal-area projection.